Doing Criminological
Research

Doing Criminological Research

Edited by
Victor Jupp, Pamela Davies,
Peter Francis

SAGE Publications
London • Thousand Oaks • New Delhi

First published 2000

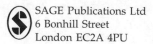 SAGE Publications Ltd
6 Bonhill Street
London EC2A 4PU

SAGE Publications Inc.
2455 Teller Road
Thousand Oaks, California 91320

SAGE Publications India Pvt Ltd
32, M-Block Market
Greater Kailash – I
New Delhi 110 048

British Library Cataloguing in Publication data

A catalogue record for this book is available
from the British Library

ISBN 0 7619 6508 4
ISBN 0 7619 6509 2 (pbk)

Library of Congress catalog record available

Typeset by Mayhew Typesetting, Rhayader, Powys
Printed in Great Britain by Redwood Books, Trowbridge, Wiltshire

To
Adam and Mark
Rory and Callum
Ann and John

CONTENTS

LIST OF FIGURES

LIST OF TABLES

LIST OF BOXES

LIST OF CONTRIBUTORS

Iain Crow has had thirty years' experience as a researcher, first at the Institute of Psychiatry's Addiction Research Unit, then as Head of Research at the National Association for the Care and Resettlement of Offenders (NACRO). In 1989 he joined the Centre for Criminological and Legal Research at the University of Sheffield. His work has included studies of the relationship between employment status and sentencing, employment and training schemes for offenders, race and criminal justice, credit and debt amongst the less well off, and the impact of the Criminal Justice Act 1991. One of his main interests has been the work of community-based organizations over a wide range of activities, and he has been an active participator in such projects as well as a researcher of them. They have included community-based projects for drug addicts, centres for the homeless, community safety projects and credit unions, as well as projects specifically directed towards offenders. He has also recently been undertaking research in youth courts, on sporting and other physical activities for young offenders, and on victim–offender mediation, and is currently writing a book on the treatment and rehabilitation of offenders.

Pamela Davies is a Senior Lecturer in Criminology and Sociology at the University of Northumbria at Newcastle. She specializes in the teaching and researching of crime and work, and crime victims. Previously she worked for Northumbria Police. Her research interests focus upon women who commit crimes for economic gain and how this fits into an informal economy. She edited *Understanding Victimisation* (Northumbria Social Science Press, 1996) as well as *Invisible Crimes* (Macmillan, 1997) (both with Peter Francis and Victor Jupp).

Jason Ditton is in the Law Department, Sheffield University, and directs the Scottish Centre for Criminology, which is at: First Floor, 19 Kelvinside Gardens East, Glasgow, G20 6BE. **Jon Bannister** is in the Department of Social Policy and Social Work, Glasgow University; **Stephen Farrall** is at the Centre for Criminological Research, Oxford University, and **Elizabeth Gilchrist** is in the School of Psychology, University of Birmingham. All four collaborated on the ESRC-funded project which generated most of the data cited in this chapter, and helped with the analysis of it presented here. Jason Ditton gave the presentation at the Annual Criminal Justice Conference held at the University of Northumbria in March 1998.

Peter Francis is a Senior Lecturer in the Division of Sociology at the University of Northumbria at Newcastle. He teaches in the area of criminology, penology and crime prevention. His books include *Prisons 2000* (with Roger Matthews, Macmillan, 1996), *Understanding Victimisation* (Northumbria Social Science Press, 1996) and *Policing Futures* (Macmillan, 1997) (both with Pamela Davies and Victor Jupp). He is on the editorial board of *Criminal Justice Matters*. He is currently writing an introductory sourcebook on criminology and a textbook on contemporary theories of crime (both forthcoming).

Jeanette Garwood is a Senior Lecturer in the Department of Behavioural Sciences at the University of Huddersfield where she teaches research methods, statistics and biopsychology. Jeanette is currently undertaking research work using trace measures as an indicator of low-level delinquency within a university library setting. She is also investigating the use of the Cognitive Interview in elicitation of criminal career information. Jeanette has worked at the Universities of Oxford, Manchester, Manchester Metropolitan, and for the Health and Safety Executive where she researched in behavioural toxicology. She has a psychology degree from University College London, a cognitive sciences degree from the University of Edinburgh and a doctorate from Green College, Oxford.

Robert G. Hollands is a Lecturer in the Department of Social Policy at Newcastle University. His primary teaching interests include youth, urban sociology and the study of popular culture and leisure. He is author of *Friday Night, Saturday Night: Youth Cultural Identification in the Post-Industrial City* (1995); *The Long Transition: Class, Culture and Youth Training* (1990) and co-editor of *Leisure, Sport and Working Class Cultures: Theory and History* (1988).

Barbara Hudson is Professor of Law at the University of Central Lancashire and was previously in the Division of Sociology, University of Northumbria at Newcastle. She teaches and researches within the fields of criminology, penology and the sociology of law, with particular interests in the impact of penal strategies on the marginalized and disadvantaged. Her publications include *Justice through Punishment: A Critique of the 'Justice Model' of Corrections* (Macmillan, 1987); *Penal Policy and Social Justice* (Macmillan, 1993); *Racism and Criminology*, edited, with Dee Cook (Sage, 1993); *Understanding Justice: An Introduction to Ideas, Perspectives and Controversies in Modern Penal Theory* (Open University Press, 1996) and *Race, Crime and Justice* (edited, Dartmouth, 1996). She has also published many articles and chapters on subjects related to the general theme of 'justice and difference'. Her current research concerns conflicts between logics of risk, prevention and justice in penal strategies and the problem of how to do justice to difference. This work draws on recent developments in criminological theory and legal philosophy as well as on developments in practice such as restorative justice, and will be published by Sage.

Gordon Hughes lectures in social sciences at the Open University. He has researched and written widely in the criminological fields of multi-agency crime prevention, 'communitarian' approaches to crime control and justice and the politics of police accountability. He has also researched and written on the future of social welfare and the regulation of public services. His recent publications include *Understanding Crime Prevention: Social Control, Risk and Late Modernity* (Open University Press, 1998), *Imagining Welfare Futures* (Routledge, 1998) and *Unsettling Welfare: The Reconstruction of Social Policy* (Routledge, 1998), which he co-edited with Gail Lewis.

Victor Jupp is Head of the Division of Sociology at the University of Northumbria at Newcastle. He specializes in the teaching of research methods and criminology. His publications include *Methods of Criminological Research* (Unwin Hyman, 1989), *Understanding Victimisation* (with Pamela Davies and Peter Francis, Northumbria Social Science Press, 1996), *Data Collection and Analysis* (with Roger Sapsford, Sage, 1996) and *Policing Futures* (with Pamela Davies and Peter Francis, Macmillan, 1997).

Carol Martin is a Research Officer at the Centre for Criminological Research, University of Oxford. She has worked within the criminal justice field in both a professional and lay capacity since 1980, undertaking responsibilities as a justice of the peace, member of a prison board of visitors, criminal justice researcher and freelance consultant. She has carried out research on community penalties and alternatives to custody for young and juvenile offenders, equal opportunities policies in the police, crime prevention and community safety, drug use and drug rehabilitation in prisons, prison boards of visitors and day visits to prison for young offenders ('Scared Straight' programmes). Her publications include two editions of *The ISTD Handbook of Community Programmes for Young and Juvenile Offenders* (Waterside Press, 1997,1998). She is currently working on a two-year Economic and Social Research Council funded research project examining conflict and violence in four prison establishments.

Roger Matthews is Professor of Sociology in the Department of Social Sciences at Middlesex University. He is co-editor with Peter Francis of *Prisons 2000* (Macmillan, 1996) and author of *Doing Time: An Introduction to the Sociology of Imprisonment* (Macmillan, 1999).

Ken Pease is Professor of Criminology at the University of Huddersfield and Director of the Applied Criminology Group. He is currently on secondment to the Home Office. He has worked on repeat victimization for 12 years and has been involved with the British Crime Survey for 14 years; his current work concerns the anticipation of crime trends. Ken has acted as a consultant to the United Nations, the Council of Europe and the Customs Cooperation Council. He is a former Parole Board member and committed supporter of the Victim Support organization. In June 1997 he was awarded the OBE for services to crime prevention.

John Pitts is Professor of Socio-Legal Studies and Joint Director of the Vauxhall Centre for the Study of Crime at the University of Luton. He is the author of *The Politics of Juvenile Crime* (Sage, 1988) and *Working with Young Offenders* (Macmillan, 1990).

Michelle Rogerson is Research Assistant at the Applied Criminology Group, University of Huddersfield. Michelle has conducted a brief evaluation of the potential contribution unobtrusive measures can make to local responsible authorities conducting crime audits. She is also conducting research to anticipate future crime trends and is investigating the role of *modus operandi* in design against crime. These projects are in collaboration with the Loss Prevention Council and the Home Office. Michelle previously worked at the data archive, University of Essex where she was involved in projects mapping the practical applications of networked access to European social science data. She studied sociology and social research methods at the Universities of Durham and Surrey.

Nick Tilley is Professor of Sociology and Director of the Crime and Social Research Unit at Nottingham Trent University. He is also a Research Consultant to the Home Office Policing and Reducing Crime Unit, within the Home Office Research, Development and Statistics Directorate. His research interests lie in crime prevention, policing, and programme evaluation methodology. He has published more than a dozen Home Office research reports, is author of *Realistic Evaluation* (with Ray Pawson, Sage, 1997), and has co-edited *Surveillance. Lighting, CCTV and Crime Control* (with Kate Painter, Criminal Justice Press, 1999).

Steve Tombs is Professor of Sociology in the Centre for Criminal Justice, Liverpool John Moores University. He is also a Director of the London-based Centre for Corporate Accountability. He has published and researched widely in the area of corporate crime, and on the management and regulation of occupational safety and health. Recent publications include, with Frank Pearce, *Toxic Capitalism: Corporate Crime and the Chemical Industry* (Aldershot. Ashgate, 1998) and, with Gary Slapper, *Corporate Crime* (London: Addison Wesley Longman, 1999).

ACKNOWLEDGEMENTS

We are grateful for the support we received for this project from all the contributors to this volume, who without exception submitted draft chapters and amended versions at speed. Our thanks go to Angela, Anne and Deborah in the Division of Sociology office at the University of Northumbria, who in one way or another have 'lived through' this project over the past year. Most of all our thanks go to Susan Doberman, who administered the project from proposal stage through to compilation and delivery of the final manuscript.

Newcastle upon Tyne, 1999

INTRODUCTION

The purpose of this textbook is to cement the relationship between criminological study and research practice. The intended audience for the book includes students pursuing research methods courses on criminology and related social science programmes, and practitioners and policy-makers who are carrying out criminological research themselves or evaluating research undertaken by others. As part of any degree programme, students will engage in research of a primary or secondary nature. They may undertake research for an essay or examination, or as part of a structured piece of independent learning including a dissertation or thesis. Similarly, doing research is central to the day to day activities of many criminal justice and related professionals, as provision under the Crime and Disorder Act 1998 has highlighted.

In order to carry out research both groups will consult a range of research methods textbooks together with criminological research studies or commentaries. However, while there are many general research methods textbooks that describe the core features of types of research design, often they do not contextualize discussions and analysis within actual studies of crime and criminal justice. Similarly, while there are many accounts of criminological research, the processes involved in its planning and doing are often neglected or confined to a short preface or appendix, while critical reflection upon the experience of doing it is nowhere to be found. It is as a response to this that this textbook has been compiled. *Doing Criminological Research* aims to bring together issues regarding the practices, strategies and principles of criminological research within the context of a range of research studies.

Research is a word which is heard more and more during everyday life, although it means quite different things to different people. The word is often used simply to mean finding out about something. Given this definition, nearly every person has been engaged in some form of research, and many of the skills used are commonplace and everyday. They include the ability to ask questions, to listen, to observe and to make notes. For example, buying a car, a computer or a piece of household furniture is a task more often than not associated with visiting a variety of stores, reading catalogues and enquiring about the possibilities of special offers, discounts and additional goods as part of the deal. When we do so, however, it does not matter much whether mistakes are made, or that the wrong information is collected, or that information is incorrectly interpreted (apart from to our own pocket or household enjoyment, recreation and comfort).

One of the ways of distinguishing scientific inquiry from such everyday inquiry is by the systematic manner in which data are collected and analysed to reach conclusions about the problem which is at the centre of the research. Research inquiry is likely to conform to specific standards, will be pursued through the use of appropriate research methods and will invariably culminate in a report of some kind, which will also conform to accepted standards. Whether 'systematic' is the same as 'scientific' in the traditional sense of the word and whether criminology can be viewed as equivalent to one of the so-called 'hard' physical sciences, however, is a matter of dispute and debate. There are lots of different ways in which criminologists go about their work. Indeed, by bringing together different criminologists within this volume, it was our intention to demonstrate the plurality of types of criminological research. In many respects the variety and form reflect the range of disciplines which comprise the social sciences. For example, the development of psychology has been influenced by the use of the experimental method to generate findings against which to test ideas. Sociology has been much more closely associated with the use of official statistics, such as published statistics on crime, as indicators of features of society and also with the use of social surveys to study large populations and social groups. Social anthropology has contributed certain types of observational methods, detailed interviews and case studies as means of producing detailed descriptions of cultures. Sometimes such methods, which are typically non-quantitative in emphasis, and which seek to understand the social world from the point of view of participants, are known as ethnographic methods. Ethnography, as applied to modern society, is used to study small scale contexts such as the courtroom or to analyse subcultures in society such as police 'canteen' culture or youth drug culture. The study of politics has many strands to it and one of the most influential involves the analysis of social structures, power and the political ideas which underpin these. This has been coupled with reflection not just on 'what is' but more especially on 'what should be' (for example in the search for a just society).

In exploring the practice of criminological research within this volume, there is an emphasis on the following four questions: How is criminological research planned in relation to what have come to be defined as important research questions? How is criminological research accomplished? What is the experience of doing criminological research in the field? What are the institutional constraints and impediments to criminological research? These questions provide the main organizing structure for the volume, and are explored in greater depth within the three main sections of the book – 'Planning Criminological Research'; 'Doing Criminological Research'; and 'Experiencing Criminological Research'.

Part I, 'Planning Criminological Research', explores the differing ways in which research problems are formulated alongside the factors which influence such formulation. It also considers the specifics of taking decisions regarding constructing research problems, the writing of proposals and the operation-alization of strategies. In Chapter 1, Victor Jupp examines the interchange between problems, theories and methods in criminological research. In doing so, he examines the role of theory in problem formulation and also as an influence on the specificity with which problems are formulated. Peter Francis in Chapter 2 illustrates the core decision-making processes associated with getting criminological research started. Drawing upon his own experiences of planning research of a community safety and crime reduction programme, his chapter describes and exemplifies various aspects of the planning process including research initiation, reviewing the literature, conceptualization, formulating research questions, project design and proposal development and presentation.

Part II, 'Doing Criminological Research', explores the many ways in which criminological research is carried out. Throughout the chapters in this section a range of themes and issues recur, including the different styles of crimino-logical research (policy-related, action-based, etc.); the different types of research design (survey, evaluative, ethnographic); and the use of different forms of data (primary and secondary; qualitative and quantitative). That said, the chapters are contextualized in differing criminological issues and are written from a variety of criminological perspectives. In Chapter 3, Steve Tombs focuses upon the problematic nature of securing reliable documentary information on the extent, nature and impact of health and safety crimes, and highlights problems in establishing the validity and therefore the legitimacy of such research in the eyes of those who fund it, while Pamela Davies in Chapter 4 examines the use of offenders and their words as sources of data in conducting criminological research. Drawing upon her own experience of doing qualitative research with women prisoners, female offenders on proba-tion and women ex-offenders, she highlights how, despite the array of methodological problems associated with this method, it can generate effective and reliable data, if conducted in a reflective manner.

During the 1980s and 1990s, and as a result of the explosion of managerial-ist thinking across the public sector, new languages and approaches have developed within the criminal justice system and related crime control organ-izations associated with measuring value for money, efficiency, effectiveness, quality and performance. These developments have impacted considerably upon the practice of doing criminological research, and, in particular, have had the effect of placing evaluation near the top of the research agenda for many government, criminal justice, voluntary and welfare agencies and organizations. In July 1998, in introducing a new £250 million crime reduction programme tackling a diverse range of crimes, the Labour government announced that the programme would be rigorously evaluated, and the find-ings used in the compilation of a national 'what works' portfolio of crime reduction initiatives. A discussion of methodological positions in the evalu-ation of criminal justice policies is a particular theme of Part II. In Chapter 5, Nick Tilley, one of the main proponents of realistic evaluation, examines in depth what is meant by it, how it differs from other evaluation methodologies and its benefits for doing criminological research. A key argument for Tilley is that realistic evaluation provides an alternative form of methodology to the quasi-experiment. In Chapter 6 Iain Crow locates and explores the strategies and the pitfalls of doing community-based criminological evaluation. Drawing upon his experience as both academic researcher and practitioner, he concludes his chapter by focusing upon his involvement in the evaluation of the Communities that Care initiative. As these two chapters highlight, there is no single view on doing criminological evaluation. The third chapter to specifically address evaluation in Part II is by Roger Matthews and John Pitts (Chapter 7). This continues and extends the theme of addressing the impact and outcomes of criminal justice policies, in the context of doing prison research. In this chapter, the focus is on devising strategies of research for evaluating the effectiveness of forms of intervention with young offenders, such as behavioural therapy and cognitive therapy.

The final two chapters in Part II highlight the many difficulties associated with using particular methods of research in specific contexts. They also express the need for innovation and reflexivity in the way in which criminologists approach data collection. In Chapter 8 Jason Ditton, Stephen Farrall, Jon Bannister and Elizabeth Gilchrist provide a critical discussion and description of social survey methodology in the context of crime victimization surveys. In doing so, they question the ability of the conventional survey method to provide valid measures of the fear of crime. A central argument of the chapter is that much of the data generated by surveys is a function of the types of questions which are asked of respondents rather than of the reality which is being probed. The discussion explores other forms of data collection such as focus groups and qualitative interviews as accompaniments to, and possibly substitutes for, large-scale surveys using standardized questions.

The need for innovation and triangulation in the choice and use of data collection methods is also the focus of Chapter 9, by Jeanette Garwood, Michelle Rogerson and Ken Pease. Focusing upon the growing concern during the 1990s with crime prevention and community safety, together with the requirements of the Crime and Disorder Act 1998 Section 6, they argue that there is a growing political emphasis on measuring the extent of crime and disorder at a local level. While in principle this may appear to be a straightforward process, Garwood, Rogerson and Pease argue that in the context of finite resources, tight deadlines regarding research completion, and the need for practical results, trace methods provide cheap and simple indices. Moreover, they argue that trace methods triangulate well. The chapter suggests that the use of everyday trace measures is justifiable not merely on methodological grounds but also because they help concentrate policy-makers' attention on the external symbols of the very real problems associated with crime and disorder which need to be addressed at the local level.

All too often textbooks only describe how, in principle, research should be planned and operationalized. However, as everyone who has done research knows, the practice and experience of research are often different. Research is a social activity which is constructed by and which influences numerous factors including the researcher, the method, the location, politics and ethics. Part III, 'Experiencing Criminological Research', offers critical reflection upon the practice of doing research. It acknowledges the issues that arise, including the relations and interactions between the researcher and the researched; those involving emotions, subjectivity and objectivity; and the many constraints on research.

In Chapter 10, Barbara Hudson explores research experience as research practice. Drawing upon a critical social theory perspective, she argues that reflection on crime and criminal justice may be viewed not solely as a personal endeavour but as a valid form of research in its own right. The chapter explores some of the research problems and issues to which such an approach might be applicable, before illustrating reflection as critical research through examination of the shift in sentencing of young people under the age of 21 who have been involved in housebreaking. Chapter 11, by Robert G. Hollands, looks at the ways in which the media report and comment upon research. Drawing upon media reportage of a study he conducted on contemporary youth and leisure activity, and in particular on the phenomenon of 'going out' in Newcastle upon Tyne, he explores the ways in which the project itself became criminalized, both in terms of the way in which research subjects were portrayed, and of how the study itself became bracketed within an overall social problem framework. In Chapter 12, Carol Martin reflects upon the practical and methodological issues surrounding the planning and operationalization of research in a prison setting. These issues include access and safety, strategies of research, power relations and conflicts of interest. The chapter provides a fitting link from the planning to the doing and experiencing of research. The final chapter of Part III, and of the book, by Gordon Hughes, examines the ways in which politics and ethical questions not merely impact upon criminological research, but also play a key role in the form criminological research agendas take. The aim of this chapter is to help the reader gain an informed and systematically sceptical understanding of the varying political and moral factors which impact on and inform contemporary criminological research.

In putting together this volume, we have provided a number of devices to facilitate the reader's learning and reflection. First, we have included at the front of each part of the book a commentary within which general thematics, specific issues and questions, alongside conceptual definitions, are 'signposted' and discussed. In some cases these commentaries direct the reader to issues which recur in different parts of the book. To develop understanding, the commentary should be read prior to the specific chapters within each part. Second, at the end of each commentary we have included a number of questions and activities. The purpose of these is to help the reader make linkages between chapters, and to develop understanding of the key issues discussed within each chapter. Third, at the end of each chapter a list of suggested readings is highlighted in boxed form. The purpose of these is to allow the reader to explore further and in greater depth the issues and points raised within each chapter. More general references can be found at the end of each chapter. Fourth, in order to facilitate the overall aim of the volume – to bring together issues regarding the practices, strategies and principles of criminological research within the context of a range of research studies – we have ensured that each of the contributors has based their writing on actual criminological research that they have been involved with 'in the field'. While not doing so in a way which is too context specific, each chapter carefully yet concisely explores methodological issues within a criminological context.

We are keen to stress the need to contextualize the practices, strategies and principles of criminological research within actual experiences in the field. Each of the chapters emphasizes critical reflection on the planning, doing and experiencing of criminological research. The volume provides discussion of the core aspects of research design and practice, together with exemplification from real research of the many issues, problems (both theoretical and practical) and questions which arise as a result of doing it. Acknowledgement of these is central to the successful doing of criminological research.

PART I

PLANNING CRIMINOLOGICAL RESEARCH

The first part of this book is concerned with planning criminological research. A central theme which runs throughout is that doing research involves engaging in a process of decision-making. One key decision concerns the choice of subject matter of research, or what is sometimes referred to as the research problem. This decision is pivotal because the research subject or problem provides the main focus for the project and is a major influence on subsequent decisions about the ways in which the project is to be accomplished.

Chapter 1 is concerned with decisions which need to be taken in relation to the formulation of criminological research problems. One obvious starting point is deciding 'what to study'. Even though a project may take many paths before its conclusion there must be some initial statement of the territory to be examined. This acts as a benchmark against which progress is measured. It also acts as a reminder to the researcher that he or she should not wander off down a fruitless cul-de-sac. One of the hallmarks of effective research is the clear formulation of research problems and questions. These guide the project by encouraging the researcher to constantly return to key issues, whilst not acting as strait-jackets to inhibit creative inquiry (and possibly reformulation of the research problem) as the project progresses. One of the hallmarks of ineffective research is a research problem which allows an investigator to lose his or her way, with the outcome that conclusions do not address what was intended.

Topics, cases, context and time

A key decision, then, concerns *topic* – what to study? This volume starts from the assumption that what can be termed the criminological enterprise is diverse, wide-ranging and, in places, fragmented. It is carried out by a variety of researchers (for example academics, policy analysts and practitioners) who work within a variety of institutions (for example, universities, central and local government, criminal justice agencies). If one wants to draw rough boundaries around a territory within which research topics and problems can be located then it can be said that criminology asks questions about the following: the nature of crime and its extent; the perpetrators of crime; victims of crime; institutions of the criminal justice system and their workings; and how each of these interacts with wider social structural dimensions such as power, inequality, social class, gender and ethnicity. Typical questions might include 'How much crime is there and how is it geographically and socially distributed?'; 'What kinds of people commit crimes?'; 'Are there any patterns to victimization in society?'; 'In what ways does the criminal justice system discriminate against categories of people?' Such questions are necessarily broad but are an essential element in the decisions about what to study. What is more, they form a platform for taking decisions about 'who to study, where and when?', that is decisions not just about *topic* but also about *cases*, *contexts* and *time*. Broad research questions can be refined and reformulated to be more incisive and penetrative to take the form, for example, 'How do urban and rural areas (context) differ in terms of victimization of racially motivated crimes (cases) in the period between 1980 and 2000 (time)?' In this way decisions are taken to open up some dimensions of a broad topic to inquiry and not others.

End purpose of research

Many factors influence such decisions, one of the most important of which is the end purpose of research. For example, where an investigator is commissioned to evaluate the introduction of some aspect of crime prevention policy, say the introduction of CCTV in shopping centres in Newcastle upon Tyne in 1999, the selection of topic, cases, context and time will typically be specified in advance by the sponsor. Even where there is a commitment to a broad academic aim of making some contribution to knowledge and to theory it will be necessary to ground empirical inquiry in specific cases, contexts and time periods. The significance of decisions about such 'grounding' lies in the limits of generalizability. That is, all research takes place in particular contexts, studying particular cases at specific times and yet aims to make broad claims beyond the particularistic scope of inquiry. The extent to which it can do so depends upon the representativeness and typicality of the contexts, cases and times which have been chosen.

Anticipating conclusions

When formulating research problems the investigator must not just consider what to study, where and when but must also anticipate the answer to the question, 'What do I want to say?' This is not to suggest that researchers can write a final report before carrying out research (although that is a possibility in some cases). Rather it is to indicate that there needs to be some anticipation of the kind of conclusion that may be reached and the kind of evidence required to support it. For example, where the aim is to evaluate the effectiveness of the introduction of some form of criminal justice policy it is necessary to formulate research problems and questions in such a way that some conclusion can be reached about such effectiveness. Chapters 5 to 7 of this volume contain discussions about ways in which policies are and should be evaluated. The main themes are first that research questions should be framed not simply in terms of whether or not a policy 'works' but in terms of how it works, with what kinds of people and in what kinds of contexts, and second that decisions regarding research design should anticipate that. There are other ways in which researchers anticipate outcome when formulating research questions. In a more radical and critical vein what is sometimes termed standpoint research seeks to pose problems and address them from a particular standpoint (for example a feminist perspective) and anticipates reaching conclusions which reflect that standpoint (see Chapter 10 in this volume). Such research is less likely to be concerned with questions about the effectiveness of specific policies and more concerned with addressing fundamental issues such as discrimination, inequality, oppression and justice.

Audiences of research

Researchers need to pose not just the question 'What do I want to say?' but also 'To whom do I want to say it?' The audiences of research findings include academic peers, policy-makers who have commissioned research, practitioners who are interested in applying findings in their work, pressure groups who want to put forward a particular viewpoint and politicians who want to formulate or justify policies. The nature of the intended audience should be anticipated when formulating research problems. This is because of the strong connection between the way in which a research problem is expressed and the types of findings and conclusions which are eventually presented. Different audiences give credibility to evidence and arguments presented in certain ways. For example, most articles in academic journals are expected to be presented in a very formal way. Further, there is a wealth of experience which indicates that policy-makers give greater credence to statistical as opposed to non-quantitative evidence whereas pressure groups often favour detailed case studies of 'deviant' cases or *causes célèbres* so as to make maximum impact. The ways in which arguments and conclusions emerge and are presented are very much influenced by earlier decisions about the nature of the research problem and how it is expressed.

Research proposals

Ultimately the aim of research is to bring forward evidence to make an argument in relation to the research problem(s). The means by which this is to be accomplished is stipulated in a research proposal, which is a statement of preliminary decisions about the ways in which such evidence will be collected, analysed and presented. A research proposal can have varying degrees of formality. In the Appendix to his classic book *The Sociological Imagination* C. Wright Mills describes the early stages of research as involving the collecting of notes, cuttings, extracts and personal thoughts. These are organized and categorized to formulate research ideas and plans, but in a manner which is constantly under review and reformation. For Mills the writing of research proposals is a continuous process of reflection and of stimulating the 'sociological imagination'. However, at the more formal end of the spectrum, grant-awarding bodies and other sponsors of research require precise written statements which address specific headings and must be submitted by a stipulated deadline. In Chapter 2 of this volume Peter Francis describes the process of writing a formal research proposal. There are variations in the context of such a proposal but typically it will address the following. First, there will be a statement about the mechanisms by which cases will be selected. Such cases may be individuals selected to be interviewed as part of a survey but they may also be documents to analyse or interactions to be observed. Second, the means by which data will be collected should be outlined. This may be, for example, by interviews, observational methods or by the use of secondary sources such as documents or official governmental statistics. Third, it is necessary to detail the ways in which data will be analysed, for example by using one or more of the computer packages which are available for this purpose. Other issues also need to be addressed, for instance time scale and budget: anticipated problems, such as gaining access to data; ethical dilemmas; confidentiality issues; and policy implications.

A research proposal is a statement of intent about the ways in which it is anticipated that the research will progress although, as most researchers will attest, the reality of how the project is actually accomplished is often somewhat different. Part III of this volume, especially Chapters 11 and 12, provides a discussion of the problems that researchers can face 'in the field'.

Validity

A primary factor in determining the content of a proposal is the research problem: an investigator will seek to design a strategy of research that will reach conclusions which are as valid as possible to the research problem. There are two aspects of validity which need to be emphasized. The first concerns whether the conclusions a researcher reaches are credible for the particular cases, context and time period under investigation. Conclusions are neither 'right' nor 'wrong'; they are more or less credible. The extent to which they are credible is the extent to which they are said to be internally valid. For example, if a researcher is investigating the effects of CCTV on levels of crime in a particular area, the strength of validity will depend on whether there is evidence that a drop in crime levels followed the introduction of cameras and also evidence that no other factor could have produced or affected the change (such as the introduction of police beat patrols).

A second aspect of validity concerns whether it is possible to generalize the conclusions to other cases, contexts and time periods. The extent to which this is possible is the extent to which conclusions are said to be externally valid. External validity is very much dependent upon the cases, contexts and time periods which form part of the research design having representativeness and typicality.

The hallmark of a sound proposal is the extent to which the research decisions which comprise it anticipate the potential threats to validity.

Typically, research proposals are sent to reviewers for some recommendation. Reviewers will be concerned with the degree of 'fit' between a research problem and the strategy proposed to investigate it. A key question concerns whether the proposed design is likely to produce valid conclusions in relation to the research problem as stated. So a clearly formulated research problem which is capable of being investigated by social scientific methods of inquiry, in a way which is as valid as possible, is pivotal in any research.

Several factors are likely to influence the degree of fit between research problem and research design and are therefore likely to affect validity. For example, decisions about research design have to be taken in the context of constraints imposed by cost and time, and there are many forms of research which cannot be justified on the grounds of ethics. Also, it is not possible to anticipate threats to validity which may occur unexpectedly and when research is under way. So all research, whether in the planning stage or in the operational stage, is a compromise between what is desirable in pursuit of validity and what is practicable in terms of cost, time, politics and ethics. This can be termed the validity 'trade-off'.

In all of this lies the value of viewing research as a form of decision-making. Focusing on decision-making at the planning stage encourages us to take decisions to rule out, as far as possible, potential threats to the validity of our conclusions. This is vital to the 'doing' of research. Focusing on decisions taken when research is under way helps us evaluate the ways in which the validity of conclusions has been affected in ways which were not – and, perhaps could not be – anticipated. This is vital to the evaluation of research which has already been completed.

SIGNPOSTS

Decision-making: the process of adjudicating between alternative courses of action and of making choices based upon such adjudication. One way of viewing the research process is as a series of decisions taken about the topic of research, the way in which research problems are formulated, the form of case selection, data collection and data analysis. The validity of research conclusions is the outcome of such decisions. Chapters 1 and 2 outline decisions taken in formulating research problems and in planning research.

Research problems: these are questions which act as a guide and a focus for research. All research begins with a research problem although this may be reformulated and refined as research progresses. Sometimes one problem may be broken down into a sub-set of secondary problems. Research problems vary in explicitness and breadth. They can comprise questions about topic, cases, contexts and time periods. See Chapter 1 for an outline of factors to consider in formulating research and also for clarification of the different meanings which are typically assigned to the terms research focus, research problem, research question and research hypothesis.

Research proposal: a statement of a research problem and set of research questions and also a detailed specification of the ways in which these will be investigated. The latter is known as research design. The contents of a research proposal can vary according to the nature and style of the research and in relation to the intended audience. Details on how to prepare a research proposal can be found in Chapter 2. A key aim of research planning is to formulate a design strategy which will generate evidence on which to base an argument and conclusion regarding the research problem. Such planning should seek to anticipate potential threats to validity.

Validity: this is concerned with the extent to which one can rely upon and trust the findings and conclusions. It involves an evaluation of the methodological objections that can be raised against the research. One aspect of validity concerns whether conclusions about the particular cases, contexts and time periods included in the research are likely to be credible. An alternative way of addressing this is by asking whether there are any other factors which could explain the findings. These are questions about the internal validity of research. A second aspect of validity concerns whether it is safe to generalize beyond the particular cases, contexts and time periods. This involves forming a judgement about the representativeness and typicality of the cases, contexts and time periods. This second aspect can be termed external validity. See Chapter 2 for a case study of planning research to maximize validity against a background of constraints such as those imposed by budget and time scale.

QUESTIONS AND ACTIVITIES

1   Read Chapter 1 and then write a sentence describing each of the following terms: research focus; research problem; research question; research hypothesis.
2   Choose a topic of criminological interest and express it in terms of research focus, research problem or question, research hypothesis.
3   Read Chapter 2 and write down the main components of a research proposal.
4   Build up a very short research proposal of your own by considering how each component could address your chosen research topic.
5   Reflect on your short proposal and ask yourself the question, 'Are there any threats to validity?'

1

FORMULATING RESEARCH PROBLEMS

Victor Jupp

Contents

The following are summaries of studies which have been carried out:[1]

Study 1: Statistics collected about fatalities at work and also about breaches of health and safety legislation were analysed in order to make some estimate of health and safety crimes. It was recognized that for a variety of reasons such statistics do not tell the full story. Therefore, the investigation also carried out a survey of victims of such crimes as well as detailed interviews with various 'whistle-blowers' on corporate crimes.

Study 2: In order to characterize the areas of a large urban conurbation, a team of investigators went to these areas and took counts of the number of damaged bus shelters and telephone boxes; the number of piles of broken glass in car parks; the amount of graffiti on walls; and the number of empty alcohol bottles found in public areas.

Study 3: Following concerns in government and police circles about the possibility that certain categories of people no longer feel safe in the communities in which they live, an investigator carried out a survey of one community, using questionnaires, to find out how many people express feelings about fear of crime.

These represent examples of social science research. Social science research is a very broad 'church' which encompasses a wide variety of methodological approaches, styles of research and practices. This variety can often be a source of dispute with regard to fundamental issues such as whether social science should adopt a natural scientific model as used in physics or chemistry; whether or not the social world can be expressed in terms of statistical models rather than qualitative descriptions; or whether social science should aim to be value-free and neutral or be committed to particular social and political standpoints. There is also variety in relation to more practical matters such as how findings should be collected, from whom they should be collected and how they should be analysed. For example, the studies summarized above make use of official statistics, detailed interviews, measures of the traces of crime and disorder, questionnaires and social surveys.

Despite this variety the studies also illustrate two commonalities which distinguish social science research from everyday inquiry. First, in each example observations were collected and assembled in a systematic way in order to reach conclusions and to put forward an argument. Second, the observations were collected and assembled, and the arguments were presented, not in a vacuum, but in relation to some research focus, problem or question. A key feature of social scientific inquiry is the centrality given to a research problem or question which has been formulated in a systematic manner.

The aim of this chapter is to characterize research problems in criminology. This is done to provide the would-be researcher with prompts to assist in the formulation of future research ideas and how they can be expressed. Effective research requires a clearly formulated research problem at its heart, one that asks the questions to which the investigator wants an answer and one that acts as a 'signpost' from the inception of a project through to the writing of a final research report. This chapter suggests that the would-be researcher needs to consider the following when formulating research problems and questions: the *purposes of research*; the *units of analysis*; *end-products of research*; *levels of specificity*; *levels of complexity*; and the *importance of meaning*. Each is considered in turn, but first there is a discussion of research problems in general and of criminological problems in particular.

Research problems

The conclusions of research will be credible and plausible only to the extent to which the questions and problems they address are clearly formulated and expressed and followed through in a consistent manner during the inquiry.

Above all, research problems and questions should be capable of being answered by some form of social inquiry. For example, the question, 'How many marriages ended in divorce in the year 2000?' can be answered by reference to official statistics. However, the question, 'Is divorce a good thing?' cannot be answered by empirical inquiry although it may be something about which one makes a moral judgement.

Each of the studies summarized at the beginning of this chapter was initiated in response to a particular problem or question and the progress of the research was guided by it. For example, the first study was concerned with answering the question: 'How many health and safety crimes are committed in the UK in any given year?' The second study addressed the question, 'What are the levels of crime and disorder in different areas of a large urban conurbation?' and the third study asked, 'How many people in the community under investigation express feelings of fear of crime?' Research questions such as these are not fixed and immutable. Indeed, they often undergo change as a project unfolds and as new dimensions open up. For example, a project which starts out investigating the number of health and safety crimes may move into a consideration of the mechanisms by which employers and other interested parties manage to make a good number of these crimes invisible. Even though questions may change, early decisions about the topic of proposed research and the way in which it is to be approached have a powerful influence on what follows. The taking of one key decision may prompt a further set of research questions. For example, the decision to ask the question, 'How many people in the community under investigation express feelings of fear of crime?' may lead to further subsidiary questions such as, 'Are there variations according to categories of people (for example, male and female, old and young) in the extent to which there is fear of crime?' and 'What is the relationship between people's fear of crime and the likelihood that they will become victims of crime?'

Criminological research problems

A first stage in formulating research problems involves deciding what to study. This can be described as deciding upon the focus of research. The research focus specifies the general type or set of social phenomena about which the research is designed to provide information and about which the investigator will reach conclusions. One way of characterizing criminological research problems is by suggesting that they emanate from a discipline or area of study which is entitled 'criminology'. It is unlikely that one would be able to achieve universal agreement on whether there is a precise set of theories or areas of study which comprise criminology or, indeed, what this set might be (see Jupp, 1996 for a discussion of this). So as to skirt around such debates and disagreements this chapter uses the phrase 'criminological enterprise' to refer to the territory within which the foci of criminological research problems are located. This enterprise is diverse, wide-ranging and, in places, fragmented. It is carried out by a variety of researchers (for example academics, policy analysts and practitioners) who work in a variety of institutions (universities, central and local government, criminal justice agencies). Criminology focuses on the following: the nature of crime and its extent; the perpetrators of crime; the victims of crime; the institutions of criminal justice and their workings; punishment and penology; and the role of the state. These broad areas provide a platform for developing research questions such as, 'Who are the perpetrators of crime and why do they do it?'; 'Who are the victims of crime and how are they socially and geographically distributed?'; 'To what extent do courts discriminate against categories of people?'; 'What role does the state play in the formulation of criminal justice policies and practices?'

Narrowing the focus

There are lots of different ways in which a broad research focus, such as an interest in victimization of crime, can lead to a narrowing of focus. One way is to ask questions about how the different 'players' in the criminal justice system relate to each other, for example, 'How do the police deal with victims of crime?' Another way is to introduce types of crime, perhaps by considering how the police deal with victims of rape in comparison with how they deal with victims of burglary. Other dimensions can be introduced to make a research focus more precise. These include types of people, types of context and a consideration of different time periods. Also, what is often referred to as critical criminology seeks to introduce wider structural dimensions such as power, inequality, social class, gender and ethnicity into research questions. Inquiry progresses by moving between levels, for example, at one level looking at rates of victimization and at another level seeking to explain the social distribution of victimization in terms of ethnicity in British society (for a more detailed discussion of critical social research see Chapter 10 in this volume and for a review of critical victimology see Mawby and Walklate, 1994 and Walklate, 1996).

Purposes of research

There are many influences on the choice of the subject for research and on the way in which it takes shape in research problems and questions (sometimes subtle distinctions are made between the terms 'research problems' and 'research questions' but for the purposes of this chapter they will be treated as one and the same). One set of influences concerns the intended purpose of research. In this regard it is possible to distinguish policy-related research; intervention-based research; theoretical research; and critical research. These distinctions are over-tidy since there can be overlaps between such types of research: an investigation which is specifically planned as an aid to policy formation can also make a substantial contribution to the development of theoretical knowledge. Nevertheless, asking questions from the outset about the intended purpose of research is important to the would-be researcher in terms of clarifying the objectives of the research. This is because research questions which address different purposes need to be expressed in different ways. For example, intervention-based research tends to be expressed in terms of the effectiveness, or otherwise, of interventions in producing intended outcomes. Theoretical research tends to be expressed in terms of searching for relationships between abstract concepts.

Policy-related research is concerned with the collection of data and the presentation of conclusions and arguments as aids to the formulation of social policies. In criminology it has been associated with the development of criminal justice policies and therefore tends to be commissioned and sponsored by central government, especially the Home Office, and by the major institutions of criminal justice, such as the police or the prison service. Sometimes it is referred to as mainstream or conventional criminology (see, for example, Cohen, 1981) and by others as administrative criminology (see, for example, Young, 1986). The research focus of policy-related investigations is largely determined by the sponsor of research with a particular emphasis on the customer–contractor relationship, that is, a relationship within which the researcher is a contractor who is required to produce certain 'deliverables' for the customer, the sponsor of the research. The deliverables will usually be specified outputs which it is thought will be an aid to decision-making about social policy. (The Home Office Research, Development and Statistics Directorate produces a regular bulletin entitled *Research Findings* which summarizes research, most of which has been commissioned and sponsored on a customer–contractor principle.)

Intervention-based research is very close to policy research in so far as it is often concerned with evaluating the success or otherwise of policy interventions, for example experimental schemes concerned with the electronic tagging of young offenders. This form of research is sometimes also known as evaluation research (see, for example, Pawson and Tilley, 1997 and the chapters by Crow, by Tilley and by Matthews and Pitts in this volume). Research questions are usually framed in a such a way as to direct the investigation to 'test' the effectiveness of policy interventions. Sponsors of research are often looking for conclusions which are expressed in terms of either success or failure of policies. This zero sum formulation is often known as the 'what works' approach to evaluation research. Recently researchers have been arguing for recognition that there may be different levels of effectiveness (short-term, intermediate and long-term); and also for recognition that policies may work in different ways in different contexts with different types of people. This is a case for 'narrowing the focus' by producing a cluster of interrelated and more specific research questions rather than one which is expressed purely in terms of 'success' or 'failure' of policy interventions. Other forms of intervention-based inquiries include action research whereby social intervention, its monitoring and action based upon such monitoring take place at the same time; and practitioner research whereby an individual practitioner such as a probation officer introduces some new working practices and then seeks to evaluate them. In this latter case, research questions are usually formulated in terms of the practices of an individual rather than general illumination.

The emphasis in *theoretical research* is on understanding and explaining human behaviour and social action, the workings of social institutions and how all of these connect with the different dimensions of social structure. Such understandings and explanations may have policy relevance but the primary aim is knowledge accumulation. The choice of research focus lies primarily in the hands of the investigator – who may be an academic or freelance researcher – and may be prompted in differing ways. A decision may be taken to research a particular area as a result of conducting a literature review and identifying gaps in knowledge (for expositions of the role of literature reviews in formulating research see Fink, 1998; Hart, 1998). For example, an undergraduate student decided to focus her dissertation on victimization of domestic violence on the early stages of pregnancy when she discovered that, despite its prevalence, not a great deal was known in the literature about its underlying causes.[2]

Ideas about future research may also result from bringing together different theoretical strands. Hollands describes how his decision to study the social life of young people in Newcastle upon Tyne emerged from his reflections on theories of youth transitions; theories of youth culture; theories relating to rapid economic restructuring; and theories of urban sociology (see Chapter 11 in this volume). Research ideas may also result from empirical inquiry. The tradition of ethnographic research includes a strand which is concerned with formulating research ideas as a result of being immersed and grounded 'in the field', perhaps by carrying out participant observation or as a result of detailed semi-structured interviews with informants (see Davies, Chapter 4 in this volume). Using empirical inquiry to generate a research focus is not confined to ethnographic qualitative work. The classic article by Selvin and Stuart (1966) describes the process of using statistical tests on large data sets derived from surveys to go 'hunting'. This involves searching through the data, without predesignation of a set of candidate variables, to look for interesting patterns worthy of further exploration.

The specific research questions which are typically asked in theoretical research vary. However, they are most likely to be expressed in ways which encourage the search for patterns in social life and for understandings and explanations of these patterns in general terms rather than in a way which addresses the effectiveness of specific policies or interventions in specific contexts and time periods.

Critical research is both theoretical and policy-related. It is theoretical in so far as it draws upon abstract concepts, such as ideology, power and discourse, and also on bodies of ideas which are expressed in terms of these. In addition, it often addresses criminal justice policies but in a critical vein rather than as a form of evaluative research which will act as an aid to management. Critical research works on the 'outside' rather than on the 'inside'. It is important to distinguish 'criticism', in its everyday usage, and 'critical analysis' as used in this context. Criticism usually refers to an evaluation which is negative, censorious or fault-finding whereas critical research refers to an analysis of forms of behaviour, of policies or of practices in terms of underlying social structural issues and theories about these. It may result in criticism of such behaviour, policies or practices, but that is not an essential requirement.

The hallmark of critical research is the movement between different levels of analysis. For example, C. Wright Mills (1970) describes the 'sociological imagination' as being concerned with explaining individual biographies in terms of their intersections with social structure and history. This would mean explaining a person's unemployment status not as a result of his or her personal attributes but as a result of structural changes in the economy and how these have historically evolved. Similarly, Barbara Hudson analyses policies and practices relating to the sentencing of young burglars in relation to wider shifts in society and in ideologies of penal policy (see Chapter 10 in this volume). The specifics of these examples vary but what they share is a movement between different levels – between the level of specific biographies and practices and the level of the structures, processes and ideologies which underpin them. The research focus of critical research, then, varies but it starts with an interest in specific actions, practices, policies or documents. It then poses research questions about the connections between these specifics and the wider and more fundamental underpinnings.

The influence of different purposes of research on the ways in which research questions are framed can be illustrated by reference to social exclusion. During the mid- to late 1990s the concept of social exclusion was placed firmly on the political agenda. A central theme was that certain groups in society were socially excluded from the mainstream by reason of factors such as lack of education, of employment and of social opportunities. The connections between social exclusion on the one hand and crime and criminality on the other were, and continue to be, made. Government social policies aimed to reduce social exclusion and thereby crime and criminality. Typically, policy-related research questions would be formulated in terms of the effectiveness or otherwise of such initiatives in particular contexts at particular times. Theoretical research would frame questions in terms of relationships between social exclusion and criminality in general and as abstract concepts. Critical research, however, would focus on the social exclusion–crime connection as a discourse which should be the focus of research in its own right. It would pose research questions such as, 'How and why has the social exclusion–crime connection come to be accepted as "knowledge" at the time that is has and with what consequences?' It might even go on to consider the role of criminological research in the production and consolidation of this connection as 'knowledge'.

Units of analysis

All research questions are framed in terms of units of analysis. These are cases such as individuals, social groupings, contexts, events, geographical areas, institutions and societies about which data are collected. They also provide a focus for the strategy of analysis. Research questions such as those in the preceding section raise important issues regarding the unit of analysis which should be used. It is possible to examine the relationship between social exclusion and crime for geographical or local authority *areas* across the country. Each area could be scored on a social exclusion index (a composite of unemployment rates, level of education, etc. for each area) and also in terms of the crime rate. Statistical measures such as correlation coefficients would provide an indication of whether there is a relationship between the two (that is, whether areas with a high score on social exclusion tend to be those with high crime rates). It is also possible to look at the relationships for *individuals* rather than for areas. In this case, statistical analysis would be carried out to establish whether individuals who score highly on a social exclusion index are also individuals who commit crimes. Both forms of analysis are equally valid for research purposes. However, what is not valid is making a direct translation of a relationship at one level, for example the geographical or local authority area, to another level, for example individuals. Put simply, just because there is a statistical relationship between a social exclusion index and the level of crime for certain areas in England and Wales it does not follow that it is socially excluded individuals who commit the crimes. The incorrectness of jumping from relationships between properties (social exclusion and crime for areas) to those same properties at another level (social exclusion and crime for individuals) is known as the ecological fallacy or the fallacy of the wrong level. It is to avoid such threats to the validity of conclusions that the unit of analysis should be clearly specified in the research questions. This places boundaries around the generalizability of conclusions. The scope of any conclusions which emerge from an investigation should be limited to the units of analysis as defined in research questions.

This general point about scope of conclusions can also apply to units of analysis *at the same level*. For example, if research is concerned with analysing decisions taken about sentencing in *one* magistrates' court it may be inappropriate to generalize conclusions to all magistrates' courts. Whether or not it is valid to do so will depend upon the extent to which the investigator can provide satisfactory evidence that the court under study is typical and representative of *all* courts. This issue about generalizability to other units of analysis at the same level is known as the problem of external validity.

End-products of research

The importance of validity serves to emphasize that in formulating research questions investigators should have some idea of the nature of the conclusion which is to be reached (but not, of course, the specific conclusions) and the kind of evidence which is required to support it. Once again, whilst recognizing the danger of undue compartmentalization it is possible to distinguish conclusions which are intended to be *descriptions*, conclusions which are put forward as *explanations* and conclusions which *evaluate policies*.

Descriptions provide information about the unit of analysis at the centre of the research question. Some descriptions are qualitative. For example, a study of decision-making in magistrates' courts typically would describe the physical layout of the court, the different roles which are played (solicitor, defendant, witness, magistrate), the type of proceedings (committal, sentencing) and the nature of decisions (legal, administrative and so on). Such qualitative descriptions are very much within the ethnographic tradition in criminological research.

Descriptions may also be numerical, for instance where a police area is characterized in terms of the number of offences recorded by the police. Areas may be described on several variables at one and the same time. The Crime and Disorder Act 1998 required local authorities and police, in liaison with other agencies and after consultation with the community, to prepare a three-year community safety strategy based upon a crime and disorder audit for each area. Such audits provide descriptions of areas using several quantitative variables. One typical audit used the following range of statistics to describe an area: crime and disorder data; probation statistics; local authority housing statistics; unemployment data; and statistics on truancy collected from schools.[3]

Typical research questions which point the way to descriptive conclusions are as follows: 'What roles are played in the courtroom?'; 'What kinds of decisions are taken by the magistrates?'; 'What is the level of crime in an area?'; 'How many children truant from school in an area?'; 'How can we describe the area in terms of crime and disorder, employment, health and education?'

Research questions geared towards reaching conclusions which offer explanations are much more likely to be expressed in terms of patterns and relationships between variables (especially in quantitative research). If an audit moved beyond characterizing an area according to a number of variables, it would be moving towards explanation. By comparing the same area over different time periods it may seek to establish whether, say, changes in levels of crime are associated with, and can be explained by, changes in levels of unemployment, quality of housing and levels of income. Research questions cast in explanatory terms, then, are more likely to be expressed in the following ways: 'What is the *relationship* between crime levels and levels of unemployment, income and housing standards?'; or, 'To what extent can changes in crime levels be *explained* by changes in levels of unemployment, income and housing standards?'

The sponsors of policy-related, especially intervention-based research are likely to expect conclusions to be expressed in evaluative terms, that is making a judgement that the policy achieved its aims. Therefore, at their simplest, evaluative research questions are usually expressed in terms of inputs and outputs with subsequent analysis geared towards before-and-after measurements of the output variable to see if there has been a significant change as a result of the input of policy. Typically, they take the form, 'Does the introduction of input X result in a change in output Y?' In some cases a research question may be more specific, hypothesizing about the degree of change in output variables, for example asking, 'To what extent does the introduction of input X result in a change in output Y?' This alternative formulation leads us to a discussion of the ways in which research questions can be posed in more specific and complex ways.

Levels of specificity and complexity

Up to now the terms research focus, research problem and research question have been used. These terms do not have any fixed technical meaning. For the purposes of this chapter the focus of research is taken to refer to the general type or set of social phenomena which is the object of investigation and about which the research is to collect findings and reach conclusions. It is the general territory which is to be surveyed and in criminological research could include topics such as fear of crime, victimization, corporate crime, penal policies, decision-making in courts or community safety programmes. 'Research problem' and 'research question' are written formulations which narrow the research focus by posing questions about the topics or object of inquiry. They should be capable of being answered in such a way as to allow conclusions to be drawn about the questions. In doing so they contain implications about the kinds of data which need to be collected, the units of analysis about which they need to be collected and the contexts in which they need to be collected. Some research questions are very narrow and explicit in this respect, perhaps grounding questions in particular contexts and in relation to particular kinds of people. Other questions are broad, merely acting as 'signposts' to the direction in which an inquiry might proceed, and may undergo constant reformulation. Research questions as signposts tend to typify qualitative, ethnographic-type research.

There is a further term, research *hypothesis*, which has a much more precise meaning. A hypothesis is a conjecture or statement about the possible relationship between two or more variables. As such it narrows the focus even more by turning research questions into testable propositions. The role of data analysis is to assist in the formation of judgements about the degree of 'fit' between data and hypothesis. This is known as hypothesis testing.

In some projects a single research question may generate many hypotheses for testing. For example, a research focus on fear of crime in the United Kingdom may lead to the research question, 'How is fear of crime socially distributed?' In turn, an investigator may generate a number of research hypotheses which suggest conjectures about the relationship between membership of social categories and expressing fear of crime. Such hypotheses are plausible implications of the research questions, based on an understanding of existing research findings and also on what sensibly might be expected. The following could be put forward for examination. The first four hypotheses (H1–H4) represent sensible and plausible lines of inquiry; the fifth hypothesis (H5) does not relate directly to the research question and probably is not worthy of investigation:

H1:   'There is a relationship between age and fear of crime.'
H2:   'There is a relationship between gender and fear of crime.'
H3:   'There is a relationship between ethnicity and fear of crime.'
H4:   'There is a relationship between previous victimization and fear of crime.'
H5:   'There is a relationship between size of toenails and fear of crime.'

Hypotheses provide a much sharper focus than general research questions and a much firmer guide to the direction a project should take in terms of the following: from whom to collect data (people of different age, gender, ethnicity and victimization histories); and what to collect data about (their fear of crime). Hypotheses also imply certain forms of data analysis, in this case statistical measures which indicate the extent to which variables are related. Yet despite providing a sharper focus, hypotheses such as the above are very basic in so far as they offer conjectures simply about relationships but say nothing about the *strength* or *direction* of the relationship. The following is a brief summary of ways in which hypotheses can be made more specific and also more complex. (The technicalities of such hypothesis formations and the forms of data analysis required to test them are beyond the remit of this chapter. For a more detailed guide consult an advanced text on social research such as Sapsford and Jupp, 1996: Chapters 9 and 10.)

Characterizing relationships

Hypotheses can move beyond relatively straightforward statements that two variables might be related to put forward conjectures that there are distinctive connections between the *values* of one variable and the values of another. For example, one could suggest that certain categories on one variable are associated with certain categories on another – H1(a): 'Women are more likely to express fear of crime than men'. Another formulation could be that high values on one variable are associated with high values on another – H4(a): 'The more times a person has been a victim of crime the higher the level of fear of crime'. This hypothesis is suggesting that the relationship between previous victimization and fear of crime is positive, in the statistical sense of the term. The opposite of this is where a negative or inverse relationship is hypothesized, whereby low values on one variable are associated with high values on another variable – H5(a): 'Short toenails are associated with higher levels of fear of crime'.

It is possible to go even further than this by expressing the relationship between two variables in numerical terms in such a way that it suggests by *how much* an increase in one variable is associated with increases in another. The form of analysis which is used to examine how well the data fit hypotheses such as these is known as regression analysis. Typically, regression analysis is based on the assumption of dependence, that is, that changes in one of the variables are dependent on changes in the other. This is important because up until now the discussion has centred around hypotheses expressed in terms of interdependence or interrelationships. However, a further way in which relationships can be characterized in hypotheses is in *causal* terms such that it is suggested that changes in one variable produce, or cause, changes in another. Such a formulation would be exemplified by the hypothesis H4(b): 'Increases in the number of times a person has been a victim of crimes *produces* increases in the level of fear of crime.' (More specific hypotheses would indicate by how much fear of crime increases for each time a person has been

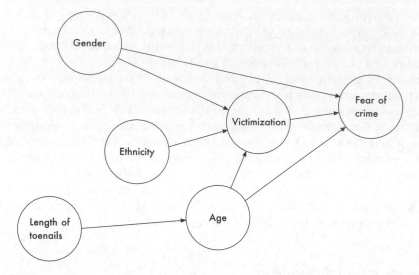

Figure 1.1   *Causal model showing hypothetical effects of five variables on levels of fear of crime*[4]

victimized.) Data analysis does not by itself produce evidence to support a causal hypothesis. In addition to statistical evidence of relationship, the researcher must be able to put forward some plausible reason as to why changes in one variable have probably brought about changes in another, and not the other way around. For example, s/he will need to explain why the more people are victimized the more they will fear crime is more plausible than the statement that greater fear of crime produces greater victimization.

Multivariate relationships

The simplest form of hypothesis is a statement about relationships between two variables and, as indicated earlier, it is possible to examine several hypotheses born out of the same research question (such as with hypotheses H1 to H5 above). However, it is highly unlikely that, say, age, gender, ethnicity, previous victimization (and length of toenails!) will have separate and independent effects on the fear of crime. To reflect this, and out of recognition of the complexity of the social world, it is possible to hypothesize about the inter-relationships between several variables and their combined and independent effects on fear of crime at one and the same time. Where such interlocking relationships and their effects are mapped it is common to refer to *modelling* social life, and the form of analysis which is used to examine the extent to which the model 'fits' the data is known as multivariate analysis (see Figure 1.1).

Advanced statistical analysis can assign numerical scores to relationships to indicate their strength, and therefore their impact on the dependent variable, in this case 'fear of crime'. This allows the model to be used for predictive purposes.

Elaborations such as these represent a high level of conjecture and analysis and are not a necessary starting point for a would-be researcher (although passing the time of day with a scrap of paper and pencil to draw lines between concepts can be a fruitful way of stimulating ideas for research). However, many would argue that even if they had the ability, the knowledge and the technology to develop and test complex hypotheses or multivariate models they would not want to do so, for several interrelated reasons.

Attending to meaning

Some of the reasons for turning away from such approaches to research are as follows. First, it can be suggested that the simplest research questions can often be the most fruitful because they do not inhibit the formulation of new ideas as a project develops.[5] Second, there is a belief that modelling distorts and imposes a false structure on social reality. The inclusion of 'length of toenails' in the model portrayed in Figure 1.1 is a reminder to us that it is often possible to find statistical relationships between variables even though there may not be any theoretical – or even sensible – reason for them being connected in reality. It is possible that 'length of toenails' is related to 'age' but not that it can have an indirect relationship, through 'age', with 'fear of crime'. This is an example of the way in which statistical models do not always correspond with social reality in a meaningful way. A third argument is that nature of social life is such that it is not amenable to statistical modelling. Fourth, it can be suggested that modelling, and also hypothesis formation, imposes preconceived ideas which are not necessarily the same as those of the people being studied, and for some researchers the subject's viewpoint is paramount. This is related to a fifth point, namely that insufficient attention is paid to the ways in which the people who are the subjects of research, experience and interpret the social world and act on the basis of such inter-pretations. It is for reasons such as these that some investigators argue that research should not begin with the formulation of conjectures in the form of precise hypotheses or causal models but with more general questions which encourage an exploration of, and progressive focusing on, the ways in which people experience and interpret their social world. In other words, research questions should pay particular attention to social meanings. What is more, they should be aimed at producing conclusions which will be grounded in such meanings and expressed in terms which would be used by the subjects of inquiry themselves.

This attention to social meanings and to deriving conclusions grounded in these meanings are two of the features of the qualitative-ethnographic tradition in the social sciences. This tradition has found expression in crim-inology in a number of ways, for example the study of subcultures of crime, the examination of interactive processes by which some people come to be labelled as criminals and others do not, and descriptions of ways in which the criminal justice system works from the viewpoint of the participants them-selves. In all such cases research questions and hypotheses are not expressed in terms of statistical relationship or causes; rather they are expressed in terms which encourage the researcher to seek explanations by understanding the social world in the terms of the subjects themselves. Research questions formulated in this way include the following: 'What kinds of stereotypes are held by police officers and how do these influence the ways in which they deal with members of the general public?'; 'What are the differing ways in which criminals are portrayed in the popular press and how do these help form popular conceptions of justice and of punishment?'; 'What labels do magistrates use when reaching decisions about the guilt or innocence of defendants?' Such research questions are aimed at producing conclusions which are based on 'explanation by understanding', that is understanding the social meanings which individuals use to interpret their world.

Ethnographic approaches can stand alone in their own right. However, they can also be used alongside more formal quantitative research. Crow (see Chapter 6 in this volume) describes the use of qualitative research in the evaluation of criminal justice policies. His argument is that evaluation research should not just be concerned with assessing the outcomes of policy interventions but also with the *processes* of interventions. Indeed, it can be argued that outcomes can only be evaluated by understanding such processes. In turn this involves 'attending to meaning' by exploring the explicit and implicit theories which sustain a programme. These may be formally expressed in the aims of the programme and in the mission statements of organizations but also include the everyday 'theories' of the participants themselves: policies do not work *on* people as objects but *through* people as subjects. In evaluating a fast tracking policy for dealing with young offenders it is necessary to formulate research questions which direct attention to the different constellations of meanings that magistrates, social workers, clients and others bring to the programme and the way in which it functions. Such questions are not fixed but are constantly reformulated as the research progresses.

Conclusion

This chapter has looked at different forms of research question and hypothesis and at influences on their formulation. In doing so it has made a number of distinctions between policy-related research, evaluative research, theoretical research and critical research. There is always a danger of over-tidy accounts of research, but the distinctions offered are not meant to provide a neat description of 'how it has been done in the past' but to prompt the would-be investigator and to assist in the development of a focus for the research and the formation of questions and hypotheses to guide it. Even though they may undergo change as an inquiry progresses, clearly formulated research questions and hypotheses are vital to giving the research direction and to ensuring that the type of conclusions which are reached are those that are intended. The following checklist should assist in the formulation of questions and hypotheses to guide research:

- What is the intended purpose of research – policy-related; action-based; accumulation of theoretical knowledge; critical theorizing? How should the purpose of research be reflected in the research question?
- What is the unit of analysis – individuals; categories of individuals; social groups; communities; institutions; societies? How should these be defined and specified in the research question? Does the intended scope of the inquiry match the units of analysis as specified in the research question?
- What are the end-products of research – description; explanation; policy recommendations? Who is the intended audience of research findings and conclusions and how should this influence the questions which direct the research?
- How specific and complex should the formulation be – questions or hypotheses-as-signposts; hypotheses specifying relationships; hypotheses specifying direction and strength of relationship; hypotheses specifying causality; multivariate models?
- Should questions and hypotheses give centrality to the ways in which the subjects of research attach meaning to the various aspects of their social life or should they reflect only the researcher's or the sponsor's focus of interest?

Finally, and perhaps most important of all, the would-be researcher should consider whether the research question is capable of being answered by methods of criminological inquiry.

Suggested readings

Bryman, A. (1988) *Quantity and Quality in Social Research*. London: Allen and Unwin.
Hart, C. (1998) *Doing a Literature Review: Releasing the Social Science Research Imagination*. London: Sage.
Jupp, V.R. (1989) *Methods of Criminological Research*. London: Allen and Unwin. Reprinted by Routledge, 1993.
Mills, C. Wright (1970) *The Sociological Imagination*. Harmondsworth: Penguin. Methodological Appendix: 'On Intellectual Craftsmanship'.
Robson, C. (1993) *Real World Research*. Oxford: Blackwell. Part 1: 'Before You Start'.

Notes

1. These summaries are based on Chapters 3, 8 and 9.
2. This dissertation entitled 'Domestic Violence During Pregnancy' was written by Gillian Goodbrand at the University of Northumbria at Newcastle in 1999.
3. This is based on the City of Durham Community Safety Partnership's Crime and Disorder Audit Report, published in December 1998.

4. This hypothetical model could be interpreted as follows: Previous victimization has a strong effect on fear of crime. In turn, victimization is influenced by gender, ethnicity and age. Ethnicity and gender interact in terms of their effects on victimization. In addition to operating through victimization, gender and age have independent effects on fear of crime. Length of toenails has no causal effect on fear of crime or on any other variable in the model. It does, however, correlate with age. Why this is the case could be the focus of a separate research question, but hopefully not within criminology.

5. 'Progressive focusing' is a term which is often used in ethnographic-type research, but not exclusively so. It refers to the continual development, refinement and perhaps redirection of research ideas in line with what is discovered to be important as the fieldwork progresses.

References

Cohen, S. (1981) 'Footprints in the sand: criminology and the sociology of deviance', in M. Fitzgerald, G. McLennan and J. Pawson (eds) *Crime and Society: Readings in Theory and History*. London: Routledge.

Fink, A. (1998) *Conducting Research Literature Reviews*. London: Sage.

Hart, C. (1998) *Doing a Literature Review: Releasing the Social Science Research Imagination*. London: Sage.

Jupp, V.R. (1996) 'The contours of criminology', in R. Sapsford (ed.) *Researching Crime and Criminal Justice*. London: The Open University. pp. 5–57.

Mawby, R. and Walklate, S. (1994) *Critical Victimology: The Victim in International Perspective*. London: Sage.

Mills, C. Wright (1970) *The Sociological Imagination*. Harmondsworth: Penguin.

Pawson, R. and Tilley, N. (1997) *Realistic Evaluation*. London: Sage.

Sapsford, R. and Jupp, V.R. (eds) (1996) *Data Collection and Analysis*. London: Sage.

Selvin, H.C. and Stuart, A. (1966) 'Data-dredging procedures in survey analysis', *American Statistician*, 20: 20–23.

Walklate, S. (1996) 'Can there be a feminist victimology?', in P. Davies, P. Francis and V. Jupp (eds) *Understanding Victimisation*. Newcastle: Northumbria Social Science Press. pp. 14–29.

Young, J. (1986) 'The failure of criminology: the need for a radical realism', in R. Matthews and J. Young (eds) *Confronting Crime*. London: Sage.

2

GETTING CRIMINOLOGICAL RESEARCH STARTED

Peter Francis

Contents

The purpose of this chapter is to describe the key decision-making processes associated with getting criminological research started. These include identifying a topic, reviewing the literature, formulating questions, making decisions about design and writing and presenting a proposal. Exemplification of these processes, drawn from my own experience of planning an evaluation of a community safety programme (Fraser and Francis, 1998; Francis and Fraser, 1998) is presented throughout the chapter in boxes, tables and figures. The use of this illustrative material brings alive key aspects of research planning and outlines proposal construction and presentation. For example, the preliminary stages associated with responding to the invitation to tender, including reviewing the literature and constructing evaluation terms of reference, are presented in box form, while the accompanying commentary describes the more specific decision-making associated with each process.

Before going on to describe the key aspects of getting criminological research started, however, three points must be highlighted. First, the processes discussed within this chapter are interrelated, inform one another and are often returned to and repeated throughout the planning process. Second, research planning involves different types of activity, from theoretical modelling through methodological problem-solving to practical decision-making, and the chapter varies in its degree of complexity, abstraction and exemplification in describing these. Third, research planning does not take place in a vacuum. Planning criminological research is a social activity and is undertaken in a social context.

The context of planning criminological research

Decisions about planning criminological research are taken within a context where politics, values and ethics influence what is researched, when, by whom, with what intention and by what means. For May (1997: 45–46), acknowledgement of this 'enables an understanding of the context in which research takes place and the influences upon it, as well as countering the tendency to see the production and design of research as a technical issue uncontaminated by political and ethical questions'.

How research problems or topics are defined is dependent upon several factors, all of which either influence values, or enable some groups' values to predominate over others (May, 1997). Equally, it is well established that criminological ideas, and with them areas of research interest, contribute to and are influenced by the wider political context (Jupp, 1989). The fact that an inquiry is worthwhile is no guarantee that it will be conducted. First, while criminological inquiry is free, carrying it out is for the most part a costly exercise. Finance and funding, and its availability and distribution, can help draw the parameters around what gets studied and how. Second, access to subjects and institutions may be contested and can be determined by subjects or gatekeepers who wish to ensure that particular practices, actions and events remain invisible or hidden (Francis et al., 1999; Jupp et al., 1999). Third, the generation of sensitive or uncompromising research can result in a general clampdown on future criminological inquiry into that area.

However, as May (1997) argues, even if topic definition were immune from politics and values, and researchers were given a free rein to undertake a project into any topic they wished, it does not follow that the planning of that research would be free of politics and values. Criminologists bring their own perspectives on the social world to the planning of criminological research, and different forms of research reflect these different standpoints (Sapsford and Jupp, 1997: 1–2). The formulation of research questions is influenced in this way, as is the choice of study population and sample. Similarly, decisions regarding the choice of method may be based as much on the criminologist's own preferences as on the applicability of the method to the task in hand. Equally, decisions concerning whether to propose a quantitative or qualitative strategy may not be based on a full assessment of their strengths and weaknesses in relation to the aims and objectives of the research. Anticipation of the needs of the sponsor, according to May (1997), can further influence the research process, including research planning.

Ethical considerations are also influential. That is, research planning should be concerned with what is right or just, in the context not only of the project, the criminologist and the sponsor, but also of the participants in the research (May, 1997). Ethical decisions depend on the politics and values of all involved in the research process and criminologists owe a duty to themselves, other researchers and their subjects and audiences to exercise responsibility at all stages. For example, according to Babbie (1998) subjects should not be forced to participate in research, research should not injure those under study, the anonymity and confidentiality of subjects should be ensured at all times, and subjects should not be deceived by what they are participating in (see also the discussion by Carol Martin in Chapter 12 regarding doing prison research). In some instances, ethical decisions are formalized as guidelines (see, for example, The British Society of Criminology Code of Ethics produced by Loraine Gelsthorpe, Roger Tarling and David Wall, January 1999 in Aizlewood and Tarling, 1999).

This brief discussion cannot do justice to the complexity of relationships between politics, values, ethics and the planning process nor has it attempted to. Its aim has been less ambitious: to indicate that values, politics and ethics can influence research planning. May (1997: 42–62) provides a thorough discussion of values and ethics in the research process, while Gordon Hughes in Chapter 13 explores the politics of research. Hall and Hall (1996) offer a good discussion of the context of planning and doing practical social research.

Having acknowledged the context of research planning, the chapter now describes the decision making involved in getting criminological research started (see also Arber, 1993a).

Defining the research topic: sponsors, criminologists and user groups

A research topic can be defined by a sponsor or can arise from a criminologist's own research activity. National and regional sponsors, including funding councils, governments and organizations of a voluntary or charitable nature with a clear interest in crime and criminal justice, such as the National Association for the Care and Resettlement of Offenders (NACRO) and Crime Concern, can define a research topic. For example, in January 1999, the Home Office's Research, Development and Statistics Directorate sought interest amongst the criminological establishment in the evaluation of restorative justice projects in England and Wales. Applicants were informed that initial expressions of interest had to take the form of a letter detailing the proposed consortium (including their relevant skills and previous experience); how the consortium might approach the research; and an explanation of why the consortium should be considered for the project (Home Office, 1999). Criminal justice agencies, such as the police and probation services, may also define a research topic, as may other local agencies with responsibility for crime, including local authorities and Single Regeneration Budget (SRB) partnerships. Often details are advertised in the local media or circulated through a tender document. An invitation to tender is presented in Box 2.1.

This invitation to tender was circulated during spring 1998. The purpose of the tender was to secure evaluation of the SRB partnership's crime reduction and community safety programme, including those initiatives focusing upon home security, the prevention of racism, trader security, CCTV surveillance, youth work, support for victims of racial harassment, truancy and exclusion, and anti-social litter and dumping. The tender is typical in that it provides information on the sponsoring organization, the geographical area in which the evaluation is to be undertaken, the background and nature of the proposed evaluation, the time frame, the evaluation brief, conditions of tender, the price for the work (omitted from the example), assessment criteria and date of submission. Other local organizations involved in research initiation include universities, where topics may be identified by supervisors for subsequent student dissertation or PhD study.

Topic identification can also arise from a criminologist's own research activity while 'in the field' or as a result of reading the literature. In some instances sponsorship may be sought to support the research. Robert G. Hollands in Chapter 11 exemplifies this process well, as do two examples from the literature. In planning research into the power elite of chief constables, Robert Reiner (1991) identified

> a number of factors which made it an attractive and interesting project. Above all, there was the growing prominence of some Chief Constables as vocal and controversial public figures . . . At the same time, a study of Chief Constables seemed a logical progression to plug a gap in the burgeoning field of police studies . . . Having previously published a study of the backgrounds, careers and occupational perspectives of the federated ranks of the police (Reiner, 1978), it seemed a logical step to attempt to conduct similar research on the elite levels. (Reiner, 1991: 39–40)

For David Nelken, defining a research topic arose from the seminal publication *Visions of Social Control* (1985), in which Stan Cohen identified the social work contract amongst other forms of control as an illustration of 'disciplinary punishment'. Critical of Cohen's bold claim, Nelken set about planning research into the extent to which Cohen's assertion was correct: 'his [Cohen's] dismissal of the value of this technique was one of the stimuli that led me to initiate this research project' (Nelken, 1989: 249).

Both means of getting research started described above are ideal typical. Most criminological research is developed through interaction and consultation between criminologists, governments, sponsoring organizations and user groups. This is the case of thematic priority funding identified by many research councils, including the Economic and Social Research Council (ESRC) (Economic and Social Research Council, 1998). Indeed, although not dedicated solely to the advancement of criminological knowledge, the ESRC has devised

Box 2.1   *An invitation to tender*

Introduction

The partnership has attracted funding through the government's Single Regeneration Budget Challenge Fund, and has embarked on a wide-ranging programme of measures. Among these are projects intended to reduce crime and the fear of crime in an area where, despite previous initiatives sponsored by the City Challenge programme, reported crime levels are still 50 per cent above the City average. The partnership wishes to commission an evaluation of the impact of its crime reduction initiatives, and a review of best practice in other areas, with a view to improving performance where feasible, targeting its resources as effectively as possible and publicizing notable successes. Tenders are invited to carry out research and evaluation, based on the following brief.

Brief for evaluation

1 To devise and implement standard measures of impact across a range of projects, covering, for example, their effectiveness in reducing crime and fear of crime, value for money, the validity of their systems for recording benefits, management arrangements, effective dissemination of results, and the consistency of methods used. This stage of the evaluation would involve recommending how better figures could be kept by all projects working in this field, so that the quality and comparability of recording and research would be improved.
2 To investigate any other special initiatives affecting crime in the area (e.g. resulting from work of the police, probation service, social service, children and young people's section, etc.) measured against the same standards.
3 To prepare an analysis of the strengths and weaknesses of existing projects and approaches, and make recommendations for improvements in practice.
4 To review experience and good practice in other areas of targeted regeneration (including any relevant initiatives in the past which have been discontinued) and to compare local initiatives with similar schemes in comparable areas.
5 To identify any gaps in provision and to make proposals for filling them, preferably by amending existing projects.
6 To identify the key measures which are likely to contribute to continued success in tackling crime and fear of crime, and make suggestions as to how they can be sustained after SRB funding ends.
7 To prepare a concise report summarizing findings and recommendations, covering both the next three years of the programme and its succession.

Conditions of tender

Tenderers are invited to submit proposals by 12 noon on Wednesday 11 March 1998. Potential consultants will be informed by Friday 20

March 1998. The evaluation is expected to be completed by 30 June
1998 with interim progress reports being presented at monthly
intervals. Tenderers are required to submit four copies of their
proposal, detailing how they would intend to carry out the work,
the names(s) and experience of those involved, and their charges.
Opportunities to work jointly with other agencies or make links
with similar research would be an advantage. An assessment will be
made on the basis of:

1 The methodology proposed to meet the terms of the brief;
2 The skills and experience of the contractor (references should
   be offered);
3 The price for the work.

Source: *Reviving the Heart of the West End* (1998) Adapted slightly
from the original.

many thematic research programmes around particular criminological issues
including 'violence' and 'crime and social order'.

Reviewing the literature and contacting stakeholders

The appropriate point at which to review the literature varies according to
different styles of research (Punch, 1998: 43). Sometimes it can lead to topic
identification. Sometimes it is undertaken at the point of topic identification,
and is related to quantitative and some forms of qualitative research, and
serves a number of functions. First, it can indicate the extent to which a topic
has already been researched. Second, it can ensure that research planning
avoids the errors of previous studies. Third, it can allow ideas to be developed
regarding good research design. Fourth, it can influence the development of a
theoretical or analytical framework. In planning a response to the invitation to
tender detailed in Box 2.1, my first task along with co-proposer Penny Fraser
of NACRO, was to review the literature. This not only refreshed and
stimulated our knowledge of the crime prevention and community safety
literature but also allowed us to examine specific issues detailed in the tender
document, such as evaluating 'value for money'. An extract from the literature
review is detailed in Box 2.2.

Additionally, reviewing the research literature helped us formulate
evaluation questions, focus the research design and finalize the evaluation
methodology. Alongside proposing both process and impact evaluation (due
to the belief that evaluation research should be concerned not just with
assessing the outcomes of policy but also with the processes of interventions –
see Iain Crow in Chapter 6), the evaluation framework subsequently

Box 2.2   Literature review

Many of the projects that are being carried out by the
partnership embody theories of intervention to prevent crime
that have been tested in other parts of the country and have
already received evaluative scrutiny. It is against this
backdrop of practice from elsewhere – and what is known about
what works, under what sets of circumstances and how – that
the projects in the area will be evaluated. For example, we
set out some national examples of successful approaches which
are reflected in the specific projects being delivered by the
community safety programme.

- **Project** – Home Security: The national evaluation of the
  first Safer Cities project by the Home Office showed that
  burglary prevention schemes were most effective when
  physical security improvements to property were
  accompanied by social measures to prevent burglary (Ekblom
  et al., 1996a, 1996b).
- **Project** – Prevention of Racism: Home Office research has
  shown that the most successful projects tackling racism
  were those that addressed the 'mutually supportive
  relationship between the individual perpetrator and the
  wider community' and 'recognized the functionality of the
  individual's expression of racism' (Sibbitt, 1997).
- **Project** – CCTV: Analysis of CCTV projects across parts of
  urban and rural Britain has identified the need to locate
  it within a package of crime prevention measures and to be
  aware of the particular ways in which it is used (Brown,
  1995). While cameras are often able to deal with
  conspicuous anti-social and criminal behaviour, evaluation
  has indicated the need for a high degree of camera
  coverage; the need for cameras to be seen playing a key
  role in the apprehension of offenders; and of the need for
  other conditions to be altered so as to improve the
  potential of CCTV in this respect.
- **Project** – Youth Work: The role played by youth work in
  crime prevention is significantly under-evaluated and this
  is partly a consequence of the reluctance of youth workers
  to regard their relationship with young people as a
  controlling one. However, recent evaluations have
  addressed this gap in knowledge (Hirschfield and Bowers,
  1997; Wiles and France 1995).
- **Project** – Support for victims of racial harassment:
  The evidence from this literature suggests that a multi-
  agency approach is essential if victims are to be
  identified, encouraged to report offences and ensured of a
  swift and appropriate response from the police and other
  agencies when victimization does occur (NACRO, 1998).
  Other research highlights the important role that can be
  played by multiracial community development work (Sibbitt,
  1997).

- **Projects** – Preventing Truancy and Exclusion: The links between persistent truancy and exclusion and crime have been widely and persuasively documented. In a study with several hundred young people by the Home Office, truancy and exclusion were found to be among the top three factors associated with offending among young people (Graham and Bowling, 1995).

Value for money

The question of value for money in crime prevention remains a vexed one. Cost benefit analyses have been limited and have been beset with a range of problems (Hough and Tilley, 1998). However, there is an accumulating body of research which is establishing principles for the evaluation of value for money. The discussion below focuses upon burglary and young people.

The savings made to the local authority, a housing association or a landlord by preventing a burglary at a property can be calculated by offsetting the cost of any preventative measures at the property against the unit cost of a burglary multiplied by the number of burglaries in the area before and after the preventative measures have been fitted. Ekblom et al. (1996a, 1996b) concluded that burglary prevention schemes funded by Safer Cities projects were cost effective in that an average burglary cost the victim and the state £1,100 whereas the average cost of preventing one burglary was about £300 in high crime areas (cited in Hough and Tilley, 1998).

The cost of a crime committed by a young person can likewise be set against the cost of an intervention to divert young people from crime. As a guide, the Coopers and Lybrand/ Prices Trust report calculated that the benefit to society of preventing a single youth crime would be the equivalent of £2,300, of which just under half would be recoverable to the public purse (Coopers and Lybrand, 1994). There are difficulties in assessing whether a particular youth diversion project has actually contributed to preventing crime, but a combination of self-report statements about offending by the young people and reports to the police of crimes where there is a known offender can be compared before and after the intervention, and an estimate of the saving can be calculated.

Some of the other projects being delivered by the partnership present a greater challenge in terms of assessing value for money. An important focus for the evaluation will therefore be to devise a means of assessing their value for money and to produce a set of recommendations for improved record keeping enabling evaluation of their value for money.

Source: Fraser and Francis (1998) Adapted slightly from the original.

proposed utilized the principles of the scientific realism school of evaluative thought (Pawson and Tilley, 1997; see also Tilley in Chapter 5). It was concerned not simply with outputs and outcomes but with the centrally important question of the linking mechanisms between these two measures of achievement. In other words the proposed evaluation was to be driven by the theory of change that informs each project (Fraser and Francis, 1998: 6).

In planning other criminological research, such as a grounded theory study (see Corbin and Strauss, 1990; Glaser and Strauss, 1967; and Strauss, 1987), reviewing the literature is delayed 'until directions emerge from the early analysis of the data. The literature is brought in later, and treated as further data for analysis' (Punch, 1998: 43). Judgements about when to review the literature are influenced by various factors, including the criminologist's knowledge, expertise and experience alongside the nature of the research topic and proposed research strategy. Yet, whilst judgements vary, the starting point for any literature review remains the library. Libraries come in different guises. They can be wide-ranging or specialized, general or academic, reference or lending. They can be a public library, a university library or attached to an organization or institution. Here it is possible to search for relevant books along with periodicals, journals, magazines, reports, newspapers and web pages using electronic and paper catalogues, together with bibliographies, abstracts, indexes, encyclopaedias (often stored on electronic databases accessed via CD ROM) and the Internet.

Accessing suitable research material can be as easy as removing it from the shelves of the library. When sources are unavailable or access to the library is limited by entrance/membership fees or borrowing restrictions, the criminologist will engage the system of inter-library loan (through which the library will request the publication from another library at a cost), visit another library or purchase the publication at a bookshop. Additionally, contact with the author of the study concerned may be made. Finally, once the material has been accessed, a critical review of the literature will be undertaken. This involves producing a thematic picture of the state of knowledge on the research topic and on the method proposed, exploring the major questions which have been investigated and considering how the literature can inform the planning of the proposed research. Hart (1998) provides an excellent discussion of doing a literature review in the social sciences.

Contacting stakeholders is another important aspect of the planning process, although the point at which to do so varies. Contacting stakeholders allows networks to be developed; ideas to be fermented; access to be negotiated; problems to be raised; and user groups to be identified. Where a research topic is defined in advance by the sponsor, forums are often organized to provide applicants with information about the topic and application process. Even where forums are not organized, the criminologist will often liaise with the sponsor and user group in order to frame the parameters of the research topic and talk through their own research thoughts and ideas. Where the criminologist is responsible for topic identification, she/ he may approach colleagues and user groups to explore further the research proposed.

Conceptualization and the formulation of research questions

A crucial aspect of research planning involves making decisions about conceptualization and the formulation of research questions. Conceptualization may be presented ahead of the research such as in planning a quantitative study, or it may be left to emerge during the research itself as in planning more qualitative research. It involves identifying the conceptual status or nature of the things to be studied and their relations to one another (Punch, 1998: 67). The formulation of research questions gives the research direction and coherence, a boundary, a focus, and a framework for writing up. It also points to the data to be collected (Punch, 1998: 38). Where research planning is informed by quantitative research logic, questions may be detailed as hypotheses that make statements about relations between variables. In other cases questions can remain relatively broad, as in more qualitative research, and are developed during the research.

In planning the evaluation of the crime reduction and community safety programme a number of factors affected conceptualization and the formulation of research questions (see Hirschfield and Bowers, 1997). The initiatives developed and implemented as part of the programme were diverse in their overall aims, intended outcomes and means of implementation, with respect to the types of input and the level of resources going into them. Specific initiatives had specific purposes: some were problem specific – for example directed at burglary reduction or racial harassment; others were geographically focused; while others again were preventable, diversionary and regulatory in nature. In delivering the programme of initiatives many agencies were working both in partnership and parallel towards a variety of aims and objectives. In order to inform our decision-making, Penny Fraser and I drew upon the work of Hirschfield and Bowers (1997: 191–192), who suggest that of most importance in formulating community safety evaluation strategies is the construction of criteria which can be used to judge the success of an initiative or programme. These authors identify a 'number of general principles and tests which evaluators might apply'. These include the extent to which the aims, objectives and targets set for each initiative are achieved according to indicators established before each initiative started; some observation and assessment of the planning, implementation and internal scheme monitoring which have been adopted; the effectiveness of inter-agency collaboration and partnership working; the strengths and weaknesses of the strategies and tactics adopted in the individual schemes; the overall impact of the policy programme or project on crime reduction and the alleviation of the fear of crime; the extent to which the projects complement and reinforce anti-crime initiatives sponsored through other policies, programmes and projects; and the extent to which individual schemes have been responsible for redistributing or displacing crime into other areas. In the light of these points, we developed research terms of reference along with identified outcomes for the purpose of planning and proposal construction, with the intention that specific questions would evolve during the research itself. The reason for this was that more information about the specific programme initiatives was needed to develop specific research questions, and this would only be available once the evaluation contract had been secured. General terms of reference and outcomes identified at the planning stage are detailed in Box 2.3.

Research design: connecting questions to data

Alongside making decisions about conceptualization and the formulation of research questions, the criminologist will also engage in the art of research design. 'Research design' is used here to refer to the stages and processes which connect research questions to data (Punch, 1998). According to Punch (1998: 66), 'When research questions are pre-specified, the design sits between the research questions and the data and it shows how the research questions will be connected to the data. When research questions are developed as the study unfolds, the design still needs to connect the questions to the data, and to fit in with both.' Research design involves making decisions about the research strategy and data collection methods, sampling, the time dimension and data processing and analysis.

At the centre of any research design is its logic or rationale – the reasoning, or the set of ideas by which any study intends to proceed in order to answer the research questions. The term 'strategy' refers to that (Punch, 1998: 67). In quantitative design, the strategy can include the experiment, the quasi-experiment and the correlation survey. While the specifics of these differ, the logic of quantitative design is more or less linear: certain stages are undertaken prior to entering the field. This linearity is detailed in Figure 2.1a and involves modelling assumed conditions and relations; exploring theoretical underpinnings; reviewing the literature and formulating pre-specified research questions or hypotheses and conceptual frameworks which are subsequently tested and perhaps falsified in empirical contexts and conditions. The logic of research is one in which theories and methods are prior to the object of research (Flick, 1998: 41).

Such linearity is associated less with the logic of qualitative research. Rather, as Figure 2.1b shows, 'there is a mutual interdependence of the single parts of the research process' (Flick, 1998: 40). In qualitative research the strategy can include the case study, ethnography and grounded theory. Grounded theory emphasizes data and context above or against theoretical assumptions. These are discovered rather than applied. Research design shows much more variability, with research questions and conceptualization often evolving more as the study progresses, while qualitative data collection methods include documents, interviews, observation and participant observation. Thus, whereas the criminologist proposing to undertake a crime survey will problematize the area under study, review the literature and detail data analysis techniques during the formative planning stages, the ethnographic criminologist will enter the field as soon as possible and undertake

Box 2.3 Evaluation terms of reference

The aims and objectives of the proposed evaluation are:

- to assess the **general** programme of measures developed and implemented within the area, detailing their

1 crime reduction and community safety effects
2 reporting, recording and monitoring arrangements
3 internal project evaluation arrangements
4 arrangements for community consultation and feedback
5 steering committee management arrangements
6 mechanisms for accountability and conformity
7 operation in the light of current thinking on 'good practice'
8 value for money

- to examine **specific** initiatives developed and implemented within the area;
- to draw upon **best practice** nationally with a view to performance improvement, effective target resource management and dissemination of good practice

The anticipated outcomes include:

- assessment of the role of key stakeholders including participating actors and agencies, along with their relations and partnerships in project management, administration and delivery;
- exploration and detailing of both shortcomings and successes;
- where possible the identification of value for money initiatives;
- identification of good recording practices by all projects so as to improve and ensure the quality and comparability of recording and research;
- identification of good practice for future developments;
- assessment of the success of the *programme* of measures, and assessment of the extent to which the stated aims of *individual* projects were realized;
- examination of the effectiveness and appropriateness of the initiatives developed.

Source: Fraser and Francis (1998) Adapted slightly from the original.

other tasks such as a literature review and conceptualization during and on completion of the fieldwork (Silverman, 1997: 2).

Sometimes, several methods are proposed in the one qualitative project, while other research designs combine quantitative and qualitative approaches. In planning the evaluation of the community safety strategy, and wishing to

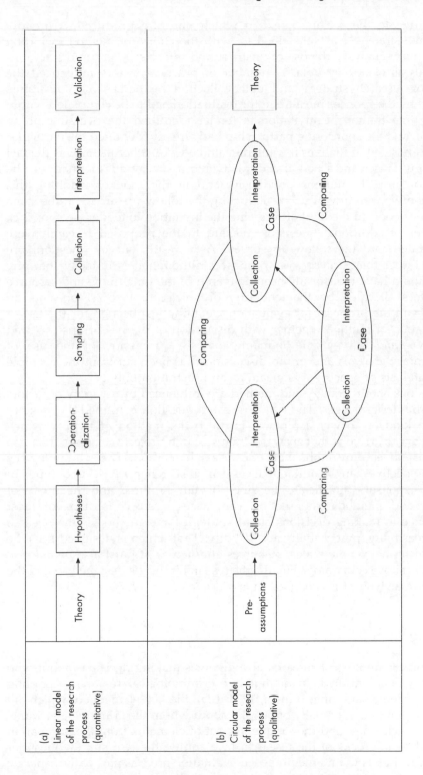

Figure 2.1   *Models of process and theory (Flick, 1998: 45)*

be sensitive to the social context in which data are produced, a broadly qualitative strategy was subscribed to, with the choice of method including documentary analysis, the use of unstructured interviews, group discussions and analysis of survey data. A number of practical factors influenced the choice of research strategy and data collection method. Time constraints imposed by the sponsor meant that the evaluation had to be completed within a short time frame; the invitation to tender identified the overall funding available; and the sponsoring partnership had already identified the particular community safety initiatives it wanted evaluated. A further factor was the fact that Penny Fraser and I had already delivered a survey questionnaire in the evaluation area 14 months previously, and at the time of planning this evaluation we were involved in conducting the follow-up survey in the same neighbourhood.[1] In the light of receiving the invitation to tender (see Box 2.1), a number of additional questions relevant to the proposed research were incorporated into the follow-up survey. As a result, survey questionnaire *analysis* rather than *delivery* was proposed, although it is unlikely that any survey could have realistically been delivered in the time frame and resource parameters allocated by the tendering organization. Box 2.4 presents an extract from the proposal for evaluation, detailing justification for the use of triangulation of method together with discussion of the use of unstructured interviews and survey questionnaire analysis (discussion of the use of documentary analysis and group discussions has been cut from the example in the interests of space – it is meant as an illustration only).

Decisions about strategy and method are influenced by a variety of factors during the design stages. The way a research question is framed dictates not only the kind of answer acceptable but also the method of inquiry which is necessary to arrive at that answer. Often research projects ask more than one question and employ more than one research method. The criminologist's own perspective can be influential, as can research purpose and type. In addition, sponsors can dictate the kinds of data required and the choice of strategy and method proposed, as can 'access issues'. Perhaps of most importance in making decisions about strategies and method is the need to ensure feasibility, practicability and validity. That which is designed must be *feasible* in terms of time scales, resources and access. It must also be *practical* within the above constraints. Finally, the research must be designed so that the conclusion arrived at is *valid* (see Sapsford and Jupp, 1997).

Sampling

Designing criminological research also involves making decisions about who or what will be studied. In quantitative criminological research, sampling refers to 'people sampling' (Punch, 1998: 105), the key elements of which are the study population (usually of people) about whom the criminologist wants to draw conclusions, and the sample, the set of elements from the population who will be the focus of the research. In planning the two survey question-naires of residents within the proposed evaluation area, as part of the separate

Box 2.4   Research methods

Triangulation of method will be used (Jupp, 1989; Sayer, 1992). This is proposed for two reasons: First, it is accepted that utilizing more that one data collection method ensures that the problems associated with one strategy may be compensated for by the strengths of another. The use of more than one method is seen as a *complementary* rather than purely *integrative* technique (Brannen, 1992): the evaluators acknowledge that merely to triangulate methods is not to eradicate all problems associated with them. It is clear that the data from different techniques do not always agree, nor do different methods always tap the same sort of information. Second, different methods are appropriate in different research situations and for collecting different types of data. A limited as opposed to multiple strategy of triangulation is proposed, involving unstructured interviews; questionnaire survey analysis; group discussions and primary and secondary documentary analysis.

Unstructured interviews

Unstructured interviews will be conducted during the evaluation with particular target groups. There are three clearly identifiable reasons for their proposed use in the current evaluation:

- Such methods allow for the collection of data which provide a detailed and full expression of informants' views.
- They are valuable as strategies for discovery   non standardized interviews can be used to find out what things are happening rather than identify the frequency of predetermined kinds of things that the researcher/evaluator already believes can happen (Loftland, 1971: 76).
- They can be administered either one-to-one or to a group.

In general terms, the benefit afforded by non-standardized interviews is their flexibility. They are often utilized to identify the main groups to be sampled; to identify mechanisms to be analysed; to lend insight into how both should be defined; and can be used to establish a variety of opinions concerning a particular topic. Group interviews in particular may be useful in assessing how several people or partnerships of people work out a common view, or a range of views about a given topic. They can help in identifying attitudes and behaviours which are considered socially unacceptable.

In the light of these points, the time frame, target group and purpose of the use of non-standardized interviews in this evaluation are detailed below. One-to-one interviews will primarily be conducted during phases one and two with local agency representatives, project workers and some community members and will serve four main aims. First, they will allow the evaluation team to get to know the various individuals involved in project development and implementation. Moreover, they will afford the evaluators an insight into the project partnership and its working relations. Second, they will allow discussion with user groups. Third, they will provide analysis of the agency members' perceptions of the projects, including their general views of project aims and objectives, and their views as to their perceived primary and secondary purpose. In addition, they will allow discussion of the perceived impact upon individuals resident or working within the various areas under evaluation and of the mechanisms through which outputs and outcomes will be secured. Finally, interviews are a very useful way of discussing avenues for continued and future developments of the project following presentation of the initial final report, taking into account best practice nationally and any local difficulties and concerns.

Survey questionnaire analysis

The evaluators have recently conducted a survey questionnaire within the area as part of a separate research project. The data will serve to provide baseline information on the programme of measures implemented. Three general reasons can be outlined for the proposed analysis of this questionnaire data. First, questionnaires are commonly accepted as an invaluable source of data about attributes, attitudes, values, personal experiences and behaviour; second they generate data in a systematic fashion (presenting all respondents with questions in a similar manner, and recording their responses in a methodical manner); and third, they address the issue of reliability of information by reducing and eliminating difference in which questions are asked and how they are presented. The survey questionnaire analysis will provide mainly quantitative data on respondents' experiences of particular projects/areas of the programme of measures; allow for an understanding of respondents' perceptions and views of the usefulness and effectiveness or otherwise of the programme/projects implemented; inform interviews conducted during phases one and two.

The complementary use of survey questionnaire and non-standardized interviews (both one-to-one and group) are seen as a professional and tried and tested methodology. Their combined use will provide the quantitative and qualitative data essential for a fully grounded evaluation of the area in question.

Source: Fraser and Francis (1998) Adapted slightly from the original.

research project detailed earlier, once the study population had been defined, probability sampling was proposed and utilized. Generally this technique involves selecting a random sample from the sampling frame, although more sophisticated approaches include cluster and stage sampling along with systematic and stratified sampling. Indeed, in designing these surveys, cluster sampling was proposed, which utilizes more than one stage of selection so as to reduce the time and costs of the research (Arber, 1993b). When the criminologist lacks a sampling frame or where a probability approach is deemed unnecessary, non-probability sampling approaches are proposed, which include purposive or judgemental, quota and snowball sampling (see Babbie, 1998 for a good discussion of sampling techniques).

Sampling is equally important in designing qualitative research, although decisions are made about settings and processes as well as populations. So the criminologist would rarely propose probability sampling but would draw upon some sort of deliberate or purposive sampling approach. For example, in designing the research of one of the community safety initiatives which provided security to local traders, snowball sampling was proposed as a way of making contact with local traders. This sampling technique operates similarly to a snowball gathering momentum (and size) when pushed along the ground. It allows the researcher to secure initial contact with a member of the sample population (in this case a local trader) who subsequently leads the researcher to other members of the same population (May, 1997). There is a variety of qualitative strategies for sampling, and their ideas vary considerably depending upon the research purpose and questions (Punch, 1998: 193).

A sampling plan is not independent of other elements of a research project (Punch, 1998: 106) and should fit the research purpose, questions, logic, form and so on. Moreover, sampling is not just part of research design. It is central to the doing of research. However, in identifying sampling and samples at the design stage, the criminologist aims to confront issues around representativeness, generalizability, reliability and validity.

The time dimension

During research design, it is necessary to acknowledge the importance of time. In planning the evaluation of the community safety initiatives, the time dimension was prescribed by the sponsor in the original tender document (see Box 2.1). Where this is not the case, the criminologist must give consideration to whether they plan to study over an extended period such as by conducting a longitudinal study or by proposing a cross-sectional study. As Babbie (1998: 104) identifies:

In designing any study you need to look at both the explicit and the implicit assumptions you are making at the time. Are you interested in describing some process that occurs over time, or are you simply going to describe what exists now? If you want to describe a process occurring over time, will you be able to make observations at different points in the process, or will you have to approximate such observations – drawing logical inferences from what you can observe now? Unless you pay attention to questions like these, you'll probably end up in trouble.

In planning criminological research it is important to acknowledge the time it will take to carry out the research, and to design a research schedule accordingly. While the form and structure of a research schedule can vary, its compilation involves mapping out the tasks to be undertaken. The schedule of research into the community safety programme was divided into three phases, with each phase accounting for particular tasks to be achieved, from getting to know the research site, through analysing the survey data and undertaking interviews to writing the final report. This is exemplified in Figure 2.2.

Data processing and analysis

Finally, in connecting research questions to data (whether quantitative or qualitative), it is necessary to make decisions about how the data are to be collected during the course of the research will be converted to a form that is immediately interpretable. Identifying the mechanisms of data processing involves making decisions about how items or groups will be assigned codes; whether written material will be altered by the addition of notes or comments; how labels will be applied to data; how significant or interesting data will be identified or chosen; and how data sets will be reduced (Blaxter et al., 1996). In proposing analysis of the survey data as part of the evaluation of the community safety initiatives, planning data processing and analysis entailed identifying the means by which raw observations detailed as ticks in boxes or words in open-ended questions would be processed including classification and computation. Electronic means of data analysis include SPSS for quantitative data and NUD*IST for qualitative data.

However, in designing for data processing and analysis, remember that while plans may seem logical and ordered, the data subsequently collected may appear anything but. It is important to allocate enough time for the collation and cleaning of data.

The research proposal

The end product for much research planning is a research proposal. A research proposal is a written document which describes the proposed research, including what it aims to do, how it will be undertaken, and the anticipated outcome. It also argues why the proposed research is important and justifies the research design, including how it connects research questions to data. In combining description and argument, a research proposal must emphasize internal cohesiveness and consistency in the planning of the research, especially where sponsorship is sought, as it is used by sponsors to make

Phase	Description of evaluation work proposed	Start Date
One	• Meet all relevant individuals and organizations involved in the programme	1 April
	• Undertake specific survey data analysis	
	• Finalize sample sizes and groups	
	• Pilot interview schedule with various target groups including various agency representatives	
	• Undertake review of best practice nationally and internationally	
	• Prepare and present preliminary evaluation findings	
	• Review experience and good practice in other areas of targeted regeneration	
	• Finalize standardized measures of impact across range of projects	
Two	• Undertake interviews with various agency representatives	20 April
	• Transcribe and analyse data from all interviews/observation	
	• Compare findings/initiatives/analysis with all relevant local primary and secondary analysis	
	• Prepare and present preliminary evaluation findings	
	• Prepare review of best practice nationally and internationally	
Three	• Undertake any further primary data collection	8 June
	• Analyse all primary and secondary data collected	
	• Write and present draft final report with recommendations for future practice	

Figure 2.2 *The research schedule (Fraser and Francis, 1998; adapted slightly)*

decisions about the nature of the research proposed and whether to grant funding. Research proposals are requirements of most research-sponsoring organizations. However, whereas funding councils and universities offering PhD studentships utilize standard forms, making planning structured and uniform, other sponsoring organizations provide limited guidelines to work within. Indeed, the invitation to tender shown in Box 2.1 detailed little in the way of guidelines on proposal construction, and the final research proposal

---

Key elements of a research proposal

---

1. Title page, title and author(s)
2. Introduction and statement of purpose
3. Research aims and objectives
4. Background literature review
5. Theoretical and practical significance
6. Research strategy and data collection methods
7. Sampling, data processing and analysis
8. Political, ethical and practical issues
9. Timetable
10. Budgetary considerations
11. Research outcomes and user groups
12. Biographical note(s) and names of referees

Figure 2.3   *The research proposal*

submission to the SRB Partnership took the form of a 17-page written document (Fraser and Francis, 1998). Where external sponsorship is not being sought it is not always necessary to write a research proposal, although many criminologists continue to do so as it allows plans to be structured, provides a written copy for review, and details evidence against which the progress of subsequent field research can be assessed. Figure 2.3 lists 12 key elements of a research proposal.

What follows is a description of these key elements. Some of these elements apply at the same time to both quantitative and qualitative research proposals; some are applicable to one more than the other. Where there are differences between the construction of primarily a quantitative and a qualitative research proposal, these are noted. In planning qualitative criminological research it is often impossible to detail what specifically will be done, so the proposal needs to explain the flexibility required, how decisions will be made as the research unfolds, and how the research will be developed. Where the chapter has not already explored in detail a particularly important aspect of proposal construction – such as budgetary considerations – an illustration from the proposal to evaluate the community safety programme is provided.

- *Title page, title and author(s)*: The front page provides the title of the study along with the names and addresses of the authors.
- *Introduction and statement of purpose*: The aim of this section is to outline the topic area, highlight the importance of the study, identify its relation to what is already known about the topic area, and provide a concise statement of purpose. The discussion may also detail how the proposed study continues work the criminologist has already been involved in; the general research strategy and whether a tightly structured or more evolving piece of research is proposed.
- *Research aims and objectives*: Under this heading conceptual frameworks are identified, together with the specific aims of the proposed research and the means by which they are to be secured – the objectives. In proposing more

qualitative research, general research statements are detailed, while an accompanying paragraph explains how conceptualization and the formulation of specific research questions will emerge during the field research.

- *Background literature review*: In quantitative research proposals, the purpose of the literature review is to place the proposed research in context; show consistency with previous research; allow for comparisons to be made; identify problems and any gaps in previous work; and provide a link to the research aims and objectives. When planning more qualitative research, the proposal will state either that the literature review will be done ahead of entering the field or that it will be integral to the research itself. In both these examples, some indication as to the nature of this literature will be given.
- *Theoretical and practical significance*: It is essential to address the question: How and why does the proposed research aim to contribute, and in what way? For more applied and evaluative designs, the proposal may also detail how the research will resolve the problems identified and whether it will recommend good practice.
- *Research strategy and data collection methods*: Discussion of the rationale for the research design together with description of the particular methods chosen is an essential part of any research proposal. It is important to show how the research methodology is appropriate and to ensure that the research design connects questions to data.
- *Sampling, data processing and analysis*: The proposed sampling frame must be outlined with discussion of how it relates to the research strategies and methods. Additionally, description and justification of procedures for data processing and analysis must be outlined, including the use of computer assisted mechanisms.
- *Political, ethical and practical issues*: Any limitations to the study which are foreseen, or any issues that have been overcome should be dealt with in this section. The proposal must show that all ethical issues have been given due consideration and reflection.
- *Timetable*: A detailed research schedule must be outlined.
- *Budgetary considerations*: Whether sponsorship is sought or not, the costs of the proposed research must be stated in the proposal, as it is surprising how quickly *hidden* costs escalate, such as the necessity to update or purchase computer software. The research budget proposed for the evaluation of the community safety programme is detailed in Table 2.1.

Budgetary categories can include staff costs (academic, management and administrative), equipment, travel and subsistence, postage and telephone, together with printing, etc. In constructing a research budget for submission to a sponsoring organization, it is important to be aware what sponsors will and will not fund. For example, employer on-costs such as central administration fees may not be forthcoming from some sponsoring organizations.

- *Research outcomes and user groups*: Sponsors often wish to know how any research findings will be disseminated to key user groups. Dissemination of research outcomes often includes the publication of reports and books

Table 2.1   *The research budget*

Item	Detailed breakdown	Cost (£)
Evaluation consultants	Peter Francis and Penny Fraser	xxxx
Stationery and related	telephone, postage, stationery, printing (of interview schedules)	xxxx
Equipment	tapes for recording interviews	xxxx
Travel and subsistence	to/from and within evaluation areas	xxxx
Printed and bound report (10)	presentation and printing of 10 copies	xxxx
Sub-total	Cumulative total without administrative charge	xxxx
On-costs	University/NACRO 10% administrative charge	xxxx
Total	Total cost of funding requested	xxxx

Source: Fraser and Francis (1998) Adapted slightly from the original.

along with the submission of articles to academic refereed journals and practitioner magazines and periodicals. However, it is often also necessary to detail how the research findings will feed back into the sphere they derived from. This is especially the case with regard to applied research and evaluation, and may involve proposing the delivery of seminars and conferences or, perhaps, focus groups with key stakeholders.

● *Biographical note(s) and names of referees*: Biographical notes of all those who have written the proposal are given at the back of the proposal, along with the names of at least two individuals who have some expertise in the research area proposed and who have agreed to act as referees.

Having written the research proposal, two additional planning stages can be identified: submission and review, and comment and revision. In seeking sponsorship, proposals are submitted and referees are appointed. The aim is to gain constructive and critical feedback, with successful sponsorship dependent upon the judgement of excellence or applicability to practice of what is proposed. In proposing to evaluate the community safety programme, Penny Fraser and I were informed by the partnership of our proposal's success within four weeks of submission. The process takes a lot longer when submission is to a funding council or national sponsoring organization. When the author of the proposal is studying for an undergraduate or postgraduate qualification in criminology, or when sponsorship is not sought, it is useful to allow a supervisor or colleague to reflect and comment upon that which is proposed. Where feedback is forthcoming, it is imperative that the proposal is revisited in the light of reviewer comments. Revision of the proposal will undoubtedly involve going back over some or all of the stages and processes discussed within this chapter, in order to ensure successful research practice.

Conclusion

This chapter has identified a number of themes. First, research planning takes place in a social context and is influenced by a variety of factors, including values, politics and ethics. Second, a sponsoring organization or a criminologist may identify the research topic, although in most cases identification of a topic arises through the interaction of sponsors, criminologists and users. Third, the planning process is often pre-structured, as in quantitative research. In other instances, such as qualitative research, it is not possible to specify exactly what will be done. Here the planning process is much more unfolding and evolving. Fourth, research design involves fitting research questions to data: what is planned must be achievable in the time frame allocated and must secure reliability and validity. Fifth, research planning is often presented as a written proposal: the aim is to describe and present an argument for that which is proposed.

Acknowledgement of these themes is important in successfully planning criminological research. Equally important is the ability to be creative, reflexive and critical. A well-planned piece of research is one which is creative, the fundamental test of which is the extent to which the proposal can solve the problem or resolve the questions detailed. Planning criminological research must also be reflexive. The most effective plans are ones where the criminologist has reflected upon all aspects from research topic initiation to proposal construction. Finally, a good research plan is one which has arisen from critical thought about the subject, about the problem, the method proposed and the ethical and political considerations attached to undertaking the research.

Suggested readings

Arber, S. (1993a) 'The research process', in N. Gilbert (ed.) *Researching Social Life*. London: Sage.
Hall, D. and Hall, I. (1996) *Practical Social Research*. Basingstoke: Macmillan.
May, T. (1997) *Social Research Issues, Methods and Process*, 2nd edition. Buckingham: Open University Press.
Pawson, R. and Tilley, N. (1997) *Realistic Evaluation*. London: Sage.
Punch, K.F. (1998) *Introduction to Social Research: Quantitative and Qualitative Approaches*. London: Sage.

Note

1. This earlier research was directed by Penny Fraser of NACRO. Peter Francis and the University of Northumbria at Newcastle managed the delivery of the survey and the computation of the data. Data analysis was undertaken by Penny Fraser.

References

Aizlewood, A. and Tarling, R. (1999) *The British Directory of Criminology*. London: ISTD and BSC.

Arber, S. (1993a) 'The research process', in N. Gilbert (ed.) *Researching Social Life*. London: Sage.

Arber, S. (1993b) 'Designing samples', in N. Gilbert (ed.) *Researching Social Life*. London: Sage.

Babbie, E. (1998) *The Practice of Social Research*, 8th edition. Belmont, CA: Wadsworth.

Blaxter, L., Hughes, C. and Tight, M. (1996) *How to Research*. Buckingham: Open University Press.

Brannen, J. (1992) *Mixing Methods: Quantitative and Qualitative Research*. Aldershot: Avebury.

Brown, B. (1995) *CCTV in Three Cities*. London: Home Office.

Cohen, S. (1985) *Visions of Social Control*. Cambridge: Polity Press.

Coopers and Lybrand (1994) *Preventative Strategy for Young People in Trouble*. London: Coopers and Lybrand.

Corbin, J. and Strauss, A.L. (1990) 'Grounded theory research: procedures, canons and evaluative criteria', *Qualitative Sociology*, 13: 3–21.

Economic and Social Research Council (1998) *Your Guide to ESRC Funding Opportunities, its Research Boards and the ESRC Office*. Swindon: ESRC.

Ekblom, P., Law, H. and Sutton, M. (1996a) *Safer Cities and Domestic Burglary*. Home Office Research Study No. 163. London: Home Office.

Ekblom, P., Law, H. and Sutton, M. (1996b) *Domestic Burglary Schemes in the Safer Cities Programme*. Home Office Research Findings No. 42. London: Home Office.

Flick, U. (1998) *An Introduction to Qualitative Research*. London: Sage.

Francis, P. and Fraser, P. (1998) *Tackling Crime and Improving Community Safety: An Evaluation for Reviving the Heart of the West End*. Newcastle and Manchester: University of Northumbria and NACRO.

Francis, P., Davies, P. and Jupp, V. (1999) 'Making visible the invisible', in P. Davies, P. Francis and V. Jupp (eds) *Invisible Crimes, Their Victims and Their Regulation*. Basingstoke: Macmillan.

Fraser, P. and Francis, P. (1998) *Reviving the Heart of the West End: Bid to Evaluate Crime Reduction Initiatives March 1998*. Manchester and Newcastle: NACRO/University of Northumbria at Newcastle.

Glaser, B.G. and Strauss, A.L. (1967) *The Discovery of Grounded Theory: Strategies for Qualitative Research*. New York: Aldine.

Graham, J. and Bowling, B. (1995) *Young People and Crime*. Home Office Research Study No. 145. London: Home Office.

Hall, D. and Hall, I. (1996) *Practical Social Research: Project Work in the Community*. Basingstoke: Macmillan.

Hart, C. (1998) *Doing a Literature Review: Releasing the Social Science Research Imagination*. London: Sage.

Hirschfield, A. and Bowers, K. (1997) 'Monitoring, measuring and mapping community safety', in A. Marlow and J. Pitts (eds) *Planning Safer Communities*. Lyme Regis: Russell House.

Home Office (1999) *Research on Restorative Justice in England and Wales: Expressions of Interest*. London: Home Office.

Hough, M. and Tilley, N. (1998) *Getting the Grease to the Squeak: Research Lessons for Crime Prevention*. Police Research Group Crime Detection and Prevention Series Paper No. 85. London: Home Office.

Jupp, V. (1989) *Methods of Criminological Research*. London: Unwin Hyman.

Jupp, V., Davies, P. and Francis, P. (eds) (1999) 'The features of invisible crimes', in P. Davies, P. Francis and V. Jupp (eds) *Invisible Crimes, Their Victims and Their Regulation*. Basingstoke: Macmillan.

Loftland, J. (1971) *Analysing Social Settings*. Belmont, CA: Wadsworth

May, T. (1997) *Social Research Issues, Methods and Process*, 2nd edition. Buckingham: Open University Press.

NACRO (1998) *Evaluation Report on the Leeds Racial Violence Surveillance Project*. London: NACRO.

Nelken, D. (1989) 'Discipline and punish: some notes on the margin', *Howard Journal*, 28 (4): 245–254.

Pawson, R. and Tilley, N. (1997) *Realistic Evaluation*. London: Sage.

Punch, K.F. (1998) *Introduction to Social Research: Quantitative and Qualitative Approaches*. London: Sage.

Reiner, R. (1978) *The Blue Coated Worker*. Cambridge: Cambridge University Press

Reiner, R. (1991) *Chief Constables*. Oxford: Oxford University Press.

Reviving the Heart of the West End (1998) *An Invitation to Tender: Evaluation of Community Safety Initiatives*. Newcastle: RHWE.

Sapsford, R. and Jupp, V. (1997) *Data Collection and Analysis*. London: Sage.

Sayer, A. (1992) *Method in Social Science: A Realist Approach*. London: Routledge.

Sibbitt, R. (1997) *The Perpetrators of Racial Harassment and Racial Violence*. Home Office Research Study No. 176. London: Home Office.

Silverman, D. (ed.) (1997) *Qualitative Research*. London: Sage.

Strauss, A.L. (1987) *Qualitative Analysis for Social Scientists*. Cambridge: Cambridge University Press.

Wiles, P. and France, A. (1995) 'Whose side are they on? Youth workers and the practice of crime control'. Unpublished paper presented at the British Criminology Conference, Loughborough.

PART II

DOING CRIMINOLOGICAL RESEARCH

How do criminologists go about their research and, in particular, how do they collect and assemble data on which to base their conclusions? Put at its simplest they can either collect data from subjects of research at first hand; they can look for physical indicators or 'traces' of patterns of behaviour; or they can use other people's data. These categories can overlap but they do allow us to consider three broad ways of working which are illustrated in this part of the volume.

Collecting data from subjects at first hand

One important method of collecting data from subjects is the social survey. It is possible to survey a wide range of units of analysis (for example interactions or documents) but typically a social survey is concerned with collecting information from individuals at first hand. This can be done either by interviewing them or by asking them to fill in a self-completion questionnaire. It is very rare to collect data from the whole of the population in which a researcher is interested: this is a very costly and time-consuming exercise. For this reason social surveys are usually sample surveys. A sample survey is a form of research design which involves collecting data from, or about, a subset of the population with a view to making inferences from, and drawing conclusions about that population. So if a researcher is interested in how many people in Glasgow are fearful of being a victim of crime it is not necessary to ask questions of all adults in that city. It is much more likely that a researcher will ask the questions of a sample of residents (say, 10 per cent) and then – using established statistical procedures – infer that the figures drawn from the sample will, with a certain degree of probability, be the figures one would have derived from the whole population. The degree to which this is likely to be the case depends a good deal on the skills of the researcher in selecting a sample which is representative of the wider population.

Social surveys have been used extensively in criminological research. One burgeoning area is the carrying out of victim surveys. In part, victim surveys developed as a result of recognized deficiencies of official crime statistics as valid measures of the extent of crime. Victim surveys collect data on the occurrence of criminal acts irrespective of whether such acts have been reported to the police and thereby form some measure of the extent of the 'dark figure' of unreported crime. Surveyors interview samples of the general population about their experience of victimization. Such surveys have been broadened to collect data on the treatment of victims by the criminal justice system, the fear of crime, attitudes to and relationships with the police. This has led to the more generic term 'crime survey'.

In Britain the passing of the Crime and Disorder Act in 1998 has acted as a spur to the completion of crime surveys based in local police areas. In Chapter 8, Jason Ditton and his colleagues describe the features of such surveys and discuss some of the issues in their implementation. Typically, large scale surveys which are concerned with aspects of crime and of victimization use structured questions as a means of collecting data from respondents. Such questions allow a researcher to present the same stimuli and thereby collect the same kinds of data from a large number of people. This not only is cheap and quick but also facilitates what is known as 'comparability of response' (which allows categories of subjects to be compared on the same items). However, such questions run the risk of being too structured. One central argument of Ditton et al. is that much of the data generated by crime surveys is a function of the types of questions which are asked rather than of the reality which is being probed. It is for this reason that they put forward an argument for the use of other, less structured, techniques for collecting data.

The use of semi-structured interviews is illustrated by Pamela Davies in Chapter 4. Semi-structured interviews involve the use of a checklist of items about which data are collected from each respondent, but they allow flexibility in the ways in which questions are asked and the order in which they are asked. They allow probing on the part of the researcher and encourage the interviewee to respond in his or her own terms and in ways which are considered significant and relevant. There are, however, disadvantages to this way of working. Semi-structured interviews are very time-consuming both in terms of data collection and data analysis and therefore cannot be used with large sample surveys. As with all social research, decision-making involves a 'trade-off'. In this case, structured interviews or self-completion questionnaires allow a small amount of information to be collected from a lot of people whereas semi-structured interviews allow a lot of information to be collected from a few people. The choice is very much dependent on the nature of the research problem; how one wants to address it; the type of audience which will receive the conclusions, and the ways in which conclusions are to be presented.

There is a further way of collecting data from or about the subjects of research at first hand, namely by evaluative methods. Evaluative methods represent the adaptation of experimental methods to an examination of the effectiveness of social policies. For example, if we are interested in the effectiveness of CCTV in reducing the fear of crime we could measure levels of fear amongst the inhabitants of two communities; then introduce CCTV into one community and not the other; finally measure levels of fear again after a suitable period of time to see whether there has been a significant reduction in the CCTV area in comparison with the other. Where the measures are taken by asking questions of samples of residents in each community the evaluative method is akin to a before–after survey. The uses of evaluative methods are illustrated in Chapters 5, 6 and 7. The main thrust of the arguments in these chapters is that the effects of social policies are much too complex to be captured by a simple before–after survey. Instead, a range of methods is needed to examine the ways in which policies work their way through different mechanisms and contexts and with differing kinds of people.

Unobtrusive measures

In the main, the methods described in the previous section have the characteristic of collecting findings from the subjects of research at first hand, typically by some form of interview or self-completion schedule. In such cases the individuals are aware that they are the subjects of research and may also be aware of the aims and objectives of the research. This introduces the strong possibility of reactive effects, or reactivity. Reactivity refers to the effect which a researcher has on what he or she is studying. For example, the subjects may react to being a part of the research and/or knowing what the research is about, perhaps by lying or exaggerating. It is for this reason that researchers sometimes turn to unobtrusive methods, that is methods which are relatively low on reactivity. They are relatively low on reactivity because, by one means or another, they remove the effect of the researcher from the interactions, events or behaviours which are being studied.

One form of unobtrusive method is participant observation whereby a researcher participates fully in the group being studied without group members knowing the true reason for the researcher's presence. Other methods involve taking – sometimes ingeniously – unobtrusive measurements of physical traces which are assumed to indicate forms and patterns of social behaviour. Such measures are sometimes also known as trace measures or behavioural by-products. Two broad types can be distinguished: erosion measures which are indicators of usage, for example the wear and tear of the most popular criminology textbooks in a library; and accretion measures, which are indicators of the deposits or residue of human behaviour, for example the number of used needles in parts of the city frequented by drug users.

In Chapter 9, Jeanette Garwood and her colleagues discuss some of the problems of crime surveys as means of measuring crime and disorder at the level of the community. These problems include the imposition of too much structure on respondents' answers, the presence of response bias as a form of reactivity and the cost of conducting surveys. This leads them to put forward the case for unobtrusive measures of crime and disorder. Such measures, they argue, are cheap and quick to collect, they reduce reactivity and they provide a clear link between what is measured and the places where community safety needs to be delivered by policy-makers, police and other practitioners.

Secondary analysis of official statistics

Secondary analysis is a form of investigation which is based upon existing sources of data and can be distinguished from primary research and analysis where the investigator collects the data for himself or herself at first hand. A secondary source is, of course, a form of accretion measure and refers to any existing source of information which has been collected by someone other than the researcher and with some purpose other than the current research problem in mind. Although there is a wide range of secondary sources available, such as reports, institutional records and memoranda, the forms of data that have a major impact on criminological research are official statistics on crime. In the UK these are recorded by the police and are notified to the Home Office. Figures are published in *Criminal Statistics for England and Wales* and in periodic *Statistical Bulletins*. There are separate publications for Scotland and for Northern Ireland. *Criminal Statistics* presents a wide range of statistics on offences recorded by the police. Statistics on recorded offences provide the 'headline' figure on the level of crime and form the basis for judgements made about society by the media, politicians, policy-makers and academics – often moral judgements about the 'social health' of society.

The deficiencies of official statistics are well known. Statistics are only collected on certain categories of crimes, those known as notifiable offences (previously referred to as 'indictable' and then, briefly, 'serious' offences), which include offences such as theft from the person, burglary and theft of and from vehicles. The police rely heavily on victims and on the general public to notify 'crime-like' acts to them. If they do not it is unlikely that the police will know about these acts let alone record them in official statistics. The gap between the true – but unknown – extent of crime and the amount of crime as measured by statistics on notifiable offences is sometimes known as the 'dark figure of crime'. Victim surveys, such as the British Crime Survey, have been developed to gain some estimate of this dark figure, of how it varies by type of offence and also to develop some understanding of people's reasons for not reporting offences to the police.

The existence of a dark figure of unrecorded crime means that the validity of official statistics on crime is questionable even where there are clear categories into which offences can be assigned, as is the case with the recording of notifiable offences. Where there is no obvious category to which a criminal act can be unambiguously assigned there can be even further question marks against validity. This issue is addressed in Chapter 3 in Steve Tombs' discussion of health and safety crimes. There are problems with the use of official crime data to make assertions about the extent of health and safety crimes in so far as their counting is often hidden in the conventional crime categories of 'notifiable offences'. Tombs turns to other forms of data: data on injuries and fatalities as recorded by the Health and Safety Executive, and enforcement data which give some indication of the incidence of actions that have been taken against employers because of their inability to comply with health and safety regulations. The problem with using the former is that it is difficult to make the leap from data on injuries and fatalities to the making of statements about crimes. The problem with the latter is that the researcher is then using data which is the outcome of the practices of enforcement agencies.

Tombs' research strategy involves finding appropriate forms of data on the basis of which to make statements about health and safety crimes and then seeking to reconstruct these in a way that allows him to ask questions about dominant assumptions and practices of employers and enforcement agencies which result in the production of health and safety data. This way of 'doing work' with its emphasis on deconstruction and reconstruction is an example not just of secondary analysis of official data but, more crucially, of critical social research. Critical social research is described more fully in Part III of this volume, especially in Chapter 10.

Issues in research

Many issues cross-cut the differing ways of going about research, as reflected in subsequent chapters. Two are worthy of note. The first concerns whether researchers are studying what they want to study. When using quantitative data this translates into the question of whether they are measuring what they want to measure. In the main, criminological researchers are concerned with concepts which are not directly observable, for example 'criminality', 'fear of crime' and 'recidivism'. However, it is possible to define a concept in such a way that the rules for making observations and for saying when an indicator of the concept has occurred can be laid down. The set of rules is known as an *operationalization*. When a researcher lays down a set of rules by which a concept can be translated into something that can be observed he or she is said to operationalize the concept. Many of the problems of planning and carrying out research are to do with ensuring that one studies and measures what is intended. As already indicated, Tombs' discussions of health and safety crimes revolve around the extent to which official crime statistics and health and safety data can be reconstituted to provide valid measures of such crimes. And in Chapter 7, Matthews and Pitts discuss whether the concept of recidivism is best measured by data on rearrests, on reconvictions, on readmissions to prison or on a combination of all three. In a different way, but addressing the same fundamental problem, Garwood and her colleagues consider whether physical trace measures such as piles of broken glass or graffiti on public buildings can be used as indicators of crime and disorder in local areas. The extent to which a researcher can devise means of observing and measuring the concepts that lie at the heart of a research problem is the extent to which there is measurement validity.

A second issue concerns the use of different methods or sources of data to study the same problem. This is sometimes known as *triangulation*. The underlying rationale for triangulation is as follows: if the same conclusions can be reached using different methods or sources of data then no peculiarity of method or of data has produced the conclusions and the confidence in their validity is increased. It is possible to distinguish different types of triangulation. Data triangulation refers to the collection of different data on the same phenomenon, for example statistics on victimization triangulated with qualitative data on the experience of victimization. Investigator triangulation involves the collection of data by more than one researcher, perhaps by the researchers adopting different roles whilst researching a prison. Method triangulation involves the collection of data by different methods which entail different threats to validity. In this way the weaknesses of one method can be 'traded off' against the strengths of another. For example, victim surveys have been used to counterbalance the deficiencies in official crime statistics by providing estimates of the so-called dark figure of crime.

Several examples of triangulation are indicated in subsequent chapters of this part of the volume. Each of the contributions on evaluative methods advocates qualitative methods to examine how criminal justice policies work and to give a more complete explanation of the efficiency of such policies than would be offered by quasi-experiments alone. Also, Tombs advocates the use of victim surveys and interviews with 'whistle-blowers' to supplement the use of health and safety data; Ditton and his colleagues make the case for focus groups as a preliminary to the use of structured questions on fear of crime; and Garwood and her colleagues treat physical traces, not as indicators in their own right, but as measures which can be triangulated with other indicators of crime and disorder.

The use of innovative ways of operationalizing concepts – so as to improve the validity of their measurement – and the triangulating of data or of method – so as to improve validity of overall conclusions – are illustrative of the need for creative decision-making in the planning and doing of criminological research.

SIGNPOSTS

Evaluative methods: these refer to methods of research which are used to assess the ways in which criminal justice policies work and their efficacy. The evaluation of policies has adapted the basic principles of the traditional experiment, which involves the formation of experimental and control groups, the introduction of some form of experimental treatment to one of these groups and measurements of the key dependent variable before and after the introduction of the 'treatment'. When applied to the analysis of social policies in real-world situations the term 'quasi-experiment' has been used. In recent years researchers have become critical of the quasi-experiment and of the search for once-and-for-all 'what works' conclusions. In their differing ways, Chapters 5, 6 and 7 argue that policies can work according to variations in the mechanisms through which they work, the contexts in which they work and the people on which they work. They also argue for a triangulation of qualitative methods with quantitative methods.

Social surveys: typically, this refers to a form of research design by which data are collected from or about a sub-sample of the population with a view to making inferences and drawing conclusions about that population (the term 'census' is generally used when all members of a population are included in a study). The cases surveyed are various but in the social sciences the survey is usually synonymous with the study of individuals and categories of individuals (their backgrounds, their attitudes and their behaviour). Data are collected either by interview or by the use of self-completion questionnaires.

The victim survey has been a particular development within criminological research. This type of survey is often used to supplement the deficiencies in official statistics on crime, especially those which result from under-reporting by the victims and the public in general. Samples of individuals are asked if they have been victims of a criminal act, whether or not they reported it to the police and if not, why not. Such surveys have also been used to estimate the number of people who fear possible victimization and to consider whether fear of crime is associated with other attributes or attitudes. They are also sometimes known as crime surveys. See Chapter 8 for a discussion of crime surveys.

Semi-structured interviews: a method of collecting data from individuals which provides a general framework for the respondents but which allows them to address issues they believe to be important, in the language which they typically use. This form of data collection is in contrast to the standardized or structured interview in which every effort is made to ask every subject exactly the same questions in the same way and in the same order. The value of semi-structured interviews is that they allow the researcher to explore in depth some aspect of the respondent's feelings, motives meanings and attitudes. Semi-structured interviews can be used to explore the dimensions of a concept such as 'fear of crime' as a precursor to operationalizing that concept in a structured interview (see Chapter 8): they can be used to collect life histories, say in the analysis of criminal careers (see Chapter 4): and they can be used to examine how individuals experience the introduction of social policies in evaluative research (see Chapters 5, 6 and 7).

Unobtrusive measures: a measure is unobtrusive if the method used to obtain it in no way affects the phenomeneon under study. This, however, is an ideal in so far as all methods have some reactive effects. The aim of unobtrusive methods, therefore, is to keep reactivity to a minimum. Sometimes they are also known as non-reactive methods.

One form of unobtrusive method is participant observation whereby the researcher participates in the groups he or she is studying, usually without the group members being aware of the true reason for his or her presence. Another form is the collection of traces left behind by forms of human action and which are taken as indicators of such action. See Chapter 9 for a discussion and illustration of these.

Secondary analysis: this refers to a form of inquiry and analysis based entirely on pre-existing data sources. It can be distinguished from primary research and analysis whereby an investigator collects the data at first hand. A secondary source is an existing source of information which has been collected by someone other than the researcher and with some purpose other than the current research problem in mind. Examples include documents such as police reports, institutional memoranda, diaries and letters. See Chapter 3 for an example of secondary analysis using health and safety statistics.

Operationalization: this refers to the laying down of rules which stipulate when instances of a concept have occurred. Operational rules link abstract concepts to observations. Such observations are sometimes also known as indicators. The extent to which observations truly indicate instances of a concept is the extent to which an operationalization has measurement validity.

Examples of operationalizations in criminological research include clusters of questions on a questionnaire to measure the concept 'fear of crime' (see Chapter 8); the use of reconviction statistics and reoffending statistics to indicate the concept 'recidivism' (see Chapter 7); and the counting of piles of broken glass in car parks to indicate levels of car crime in an area (see Chapter 9).

Triangulation: this is the use of different research methods or sources of data to examine the same problem. It derives its rationale from surveying, in which one locates an objective or a building by 'fixing' it in relation to two or more places. Triangulation can be used at different stages of the research and for different purposes. For example, qualitative methods such as focus groups may be used prior to structured interviews to open up the varying dimensions of a concept, such as fear of crime (see Chapter 8). Trace measures may be used alongside crime surveys to validate the measurement of crime and disorder in an area (see Chapter 9), and semi-structured interviews can be used alongside quasi-experimental designs to understand how social policies are experienced by subjects in an experimental programme (see Chapters 5, 6 and 7).

QUESTIONS AND ACTIVITIES

1   After reading Chapter 3 plan a strategy for measuring the extent of health and safety crimes in your area.

2   Read Chapters 5, 6 and 7 and then answer the following:
    (a)   What are the critiques of the traditional quasi-experimental (OXO) model?
    (b)   What methods can be used to study the ways in which policy interventions are experienced by participants?

3   Read Chapter 8 and then prepare a short questionnaire, comprising no more than four questions, which you believe will operationalize the concept 'fear of crime'. To what extent do you think your operationalization is valid?

4   Compare the strengths and weaknesses of semi-structured interviews (see Chapter 4) with structured interviews (see Chapter 8).

5   Read Chapter 9 and devise some trace measures to estimate the extent of each of the following in your community: under-age drinking in public areas; racism; vandalism. To what extent do you think your operationalizations are valid?

3

OFFICIAL STATISTICS AND HIDDEN CRIME: RESEARCHING SAFETY CRIMES

Steve Tombs

Contents

Safety crimes and official crime statistics

While the limitations of 'official crime statistics' are now well documented, these remain predominant as a 'descriptive medium' (Maguire, 1997: 139) in crime debates, constituting at least a starting point for almost all efforts to quantify the extent of various forms of offences. The particular concern of this chapter is with the availability – or otherwise – of official data on safety crimes. Safety crimes are infractions of a legal duty placed upon an employer or a corporate entity as defined within the framework of the criminal law, most notably the Health and Safety at Work Act (HASAW Act) 1974.

Even a cursory examination of 'notifiable offences' for England and Wales reveals the extent to which the focus is almost exclusively upon 'conventional' crimes. The most likely categories for coverage of safety crimes within this data are those relating to homicide, manslaughter and violent crime. In fact, the crime of homicide has been legally and socially constructed in a way that renders it inapplicable to corporate offences/offenders, not least because of the centrality of *mens rea* to this offence (Wells, 1993). Perhaps a more fruitful source of data on safety crimes is the offence category of manslaughter. The problems in attempting to apply this charge successfully to both the corporate entity and/or individual employers are well known (see Slapper and Tombs, 1999); and given that, at the time of writing, just four successful manslaughter prosecutions for safety offences have been recorded, such data are unlikely to make much of an impact in the relevant Home Office's data column. Finally, the category of violent crime – in particular the sub-category of 'violence against the person' – might also legitimately include reference to infractions of the criminal law which resulted in injury (fatal or otherwise) to an employee or member of the public. Again, this is not actually the case. One of the most striking aspects of legal – not to mention political and academic – treatments of violence is the general absence of the effects of corporate activities from these (Levi, 1997; Pearce and Tombs, 1992; Wells, 1993, 1995).

One of the underlying assumptions of this chapter is that there are overwhelming reasons why safety crimes – employer infractions of the criminal law, which result in deaths or serious injury – should be considered as crimes of violence (Pearce and Tombs, 1998; Slapper and Tombs, 1999). While safety crimes represent both a social and a 'law and order' problem (Tombs, 1999), their relative absence from treatments of violence reflects something of a self-perpetuating process: given the absence of utilizable data on safety crimes, it is difficult to intervene in crime debates on the issue of safety crimes, while the relative lack of attention that is paid to them by (and, indeed, beyond) criminology means that efforts in, and progress on, the production of utilizable quantitative data on safety crimes is slow and uneven, to say the least. A minimal requirement for critical social science is to question existing dominant, including official, assumptions and the practices (not least the collection and categorization of data) which follow from these. Thus one necessary response to the absence of data on safety crimes in official crime statistics is to seek other sources of such data and, if necessary, reconstruct them. This latter aim is the focus of this chapter.

Safety crimes and official data: measuring injuries

If data on safety crimes are not held by the key bodies responsible for recording and reproducing *crime* data – the police and the Home Office – then the next port of call is that of data held by the government agency responsible for recording and reproducing occupational *safety* data. The relevant body here is the Health and Safety Executive (HSE), an umbrella agency established under the HASAW Act 1974, into which existing inspectorates were relocated, and which is responsible for enforcing the Act and all related legislation.[1]

Injury data

An obvious starting point here are data collated under the Reporting of Injuries, Disease and Dangerous Occurrences Regulations 1995 (RIDDOR 1995). It is important to be clear, however, that in the context of a treatment of safety crimes, RIDDOR can provide raw material only: that is, RIDDOR requires the reporting of data on *injuries*, rather than *crimes*, a distinction that is not made in the otherwise exemplary work of both Box (1983) and Reiman (1979) on safety crimes, each of which falls foul of equating injury data with safety offences. We need to establish utilizable quantitative data on injuries prior to a consideration of what such data might tell us about safety crimes. Ideally, such an exercise would draw upon each of the categories of injury data maintained under RIDDOR, that is, 'fatal', 'major' (including amputations, most fractures and dislocations, loss of sight and some burns) and over-three-day injuries, where 'incapacity for normal work [lasts] for more than 3 days'.

Each category of *non-fatal* injury data maintained under RIDDOR is subject to significant under-reporting, a phenomenon highlighted over a quarter of a century ago by the report of the Robens Committee on Safety and Health at Work (Robens, 1972: 135, and Chapter 15, *passim*). Although major injury data are more reliable than data for over-three-day injuries, each is subject to under-reporting, which renders them somewhat unreliable as measures of occupational injury (see Nichols, 1997). It has now been estimated that official data record on average only about 40 per cent of injuries which are notifiable to HSE, and just 10 per cent of injuries to the self-employed (HSC, 1997: 68–79; Stevens, 1992); reporting rates vary between 11 per cent in 'Finance and Business' and 17 per cent in 'Hotels and Restaurants' to 62 per cent in 'Public Administration and Defence' and 86 per cent in 'Extraction and Utility Supply'. Using the expanded reporting requirements under RIDDOR 1995 (see pp. 67–8 below), data indicate that 'only one quarter of reportable non-fatal injuries to employees' and 'less than 5%' in the case of self-employed workers are reported by employers (HSC, 1998: 1). Moreover, given shifts in levels of non-reporting, these levels of underestimate render any longitudinal analysis virtually impossible – a task already made difficult by the fact that there have been several changes in reporting requirements since 1974 (see Nichols, 1997). Given the recognized 'poverty' of minor and major injury data, the variable social processes out of which they are produced, and the shifting legal categories into which they are organized, they are simply too unreliable for research purposes.

Fatality data

If there are enormous methodological and technical problems in determining accurate figures for the numbers of occupational injuries, one consistent claim has been that the recording of *fatal* injuries is at least reliable (HSC, 1996b: 1; 1997: 1; Nichols, 1997: 126; 1989: 543; 1994: 104). Indeed, the data on fatal injuries *are* the most reliable available data on occupational injury, perhaps even having been improved recently with the introduction of new reporting requirements.

Data for the reporting year 1996–97 are the first to be collected under new reporting requirements, RIDDOR 1995. Of interest is that the new regulations had expanded the definition of occupational fatalities, including violence against employees and new categories of reportable occupational fatalities amongst members of the public, by introducing the 'vital' test of 'arising out of or in connection with work' (HSE, 1996: 14) in determining what injuries should be recorded as occupational. This phrase is intended to cover 'the manner or conduct of an undertaking', 'the plant or substances used for the purposes of the undertaking', and 'the condition of the premises used by the undertaking or of any part of them' (HSE, 1996: 14). This new, more inclusive, definition simply renders reporting requirements consistent with the substance of the HASAW Act 1974, which requires an employer to 'conduct his undertaking, in such a way as to ensure, so far as is reasonably practicable, that persons not in his employment, but who may be affected, are not exposed to risks to their health and safety'. Under this expanded definition, a total of 696 occupational fatal injuries were recorded for 1996–97.

Unfortunately, while fatal injury data are clearly more utilizable than other injury data, and while the requirements of RIDDOR 1995 are an improvement, there remain at least three separate problems with this data which render its use problematic, and claims regarding its 'virtual' completeness simply wrong.

First, the coverage of RIDDOR 1995 continues to exclude significant numbers of occupationally caused deaths, for which there do exist officially collected data. Some of these omissions are at the very least curious (Tombs, 1999a), but the least excusable, and numerically most significant, is the exclusion of road traffic accidents where fatal injuries involve 'at work' vehicles. RoSPA (1998) has estimated that there are 800–1000 such fatalities per annum (see also Bibbings, 1996). Incorporating existent, omitted data increases the total of occupational deaths from the HSE's expanded figure of 696 for 1996–97 to somewhere between 1,600 and 1,900 (see, for a more detailed treatment, Tombs, 1999, and Tombs, 1999a).

Second, the data maintained under RIDDOR with respect to occupational fatalities remain internally incoherent. Anomalies arise, in particular, from inconsistent reporting requirements depending upon where a fatality occurs; the time between the injury sustained and death occurring (and here there is a disjunction between the regulatory requirement and its actual operation); and whether the fatal injury is sustained by an employee, a self-employed worker or a member of the public. The net effect is to omit significant groups, but unquantifiable numbers, of relevant fatal injuries (Tombs, 1999a).

Third, and finally, RIDDOR reflects an under-recording of fatal injuries due to social processes of under-reporting.[2] Before a death can be registered, a valid certificate giving the cause of death must be completed and signed by a registered medical practitioner who attended the deceased during her/his last illness; if a death is shown to be 'violent or unnatural', the Coroner is required by law to conduct an inquest. Two recent studies (Start et al., 1993, 1995) highlighted the inability of both hospital clinicians and general practitioners to recognize some categories of reportable deaths. The first (Start et al., 1993) indicated that individual clinicians at all grades showed a variable appreciation of the different categories of cases which should be reported, being mistaken in up to 60 per cent of individual cases. Indeed, 'Deaths resulting from accidents were often unrecognized' (Start et al., 1993: 1039). In a second study involving general practitioners (Start et al., 1995), only 3 per cent recognized all those deaths which should be reported for further investigation; deaths from industrial or domestic accidents were recognized as cases requiring referral by fewer than half of general practitioners. Both studies indicate that certifying doctors tend to consider only the eventual cause of death rather than the sequence of events leading to death (Start et al., 1995: 193).

Each of the these limitations renders problematic the use of fatal injury data as raw material for attempts to quantify safety crimes. Indeed, it is almost impossible to state at present how many people die as a result of occupational injuries in Britain each year: while HSE's 'headline' figure includes at best some 20 per cent of recorded occupational deaths, the effect of existent legal categories and other social processes produces an indeterminate under-estimation of the numbers of such fatalities (Tombs, 1999a).

Several conclusions offer themselves following this consideration of the use of fatality data maintained under RIDDOR 1995 as a basis from which to measure the extent of safety crimes. First, while fatality data are inadequate, they are liable to some reconstruction by resort to other, official measures of occupationally caused fatalities; this allows reference to what should be called the *minimal* number of occupationally caused deaths in Britain on an annual basis. Second, it remains clear that inconsistencies within the reporting requirements themselves, along with various social processes of under-reporting, mean that any reconstructed figure remains an underestimate. Third, reference to even a reconstructed total of occupational fatalities still leaves us in the realms of *injuries* rather than *crimes*. Finally, even if fatality data *can* be used to say something about safety crimes this would still exclude the vast majority of safety crimes, since crimes which result in fatalities represent only a small minority of such offences.

Fatal injuries as crimes?

It is possible, however, to say something about the proportion of these fatal injuries which might be the result of safety crimes, although this must be done in the absence of any reference to formal enforcement activity. Three sources of evidence are available which allow some – albeit tentative – conclusions on the volume of this most egregious form of safety crime: a series of special investigations into groups of fatalities undertaken by HSE in the 1980s; a now considerable stock of detailed material within reports which resulted from commissions of inquiry into a whole series of disasters that occurred in the UK; and a series of (HSE) accident investigation reports, comprising one-off but highly detailed examinations of individual incidents, usually resulting in one or a small number of fatalities. These have been discussed elsewhere (Tombs, 1999); suffice to say here that all three attempt to assign responsibility for injuries and incidents, and also the focus of each is mainly on fatalities. In all cases, the data presented tend to the conclusion that in the clear majority of cases of workplace fatalities – in at least two out of three fatal injuries – there is at least *prima facie* evidence of violations of duties placed upon employers by the HASAW Act, and thus at the very least a *criminal* case to answer. This general conclusion – based upon evidence indicating attribution of responsibility for the fatalities in question – also holds in the case of deaths at work whilst driving, with employers failing to meet their legal duties to provide safe systems of work, and to reduce risks 'so far as is reasonably practicable', where they are requiring employees to drive as part of their employment. Key causal factors in fatalities include the failure to consider safer, alternative means of transport or indeed routes; the setting of unsafe schedules, journey times and distances; failure to maintain vehicles adequately; failure to invest in vehicles with additional safety features; and the lack of specialized training on offer for regular drivers (RoSPA, 1998; see also Slapper and Tombs, 1999: Chapter 4).

Thus while there are enormous problems in determining the number of occupational fatal injuries, and in determining how many of these result from criminal activity or omission on the part of employers, it is absolutely certain that *the scale of deaths from safety crimes far outnumbers deaths from homicide* (Tombs, 1999). Equally clear is that the inability to quantify these crimes renders this argument less forceful than it might otherwise be.

Safety crimes and official data: measuring offences

A second potential source of data regarding safety crimes is data on enforcement activity. One advantage of exploring enforcement data provided by HSE is that these are an index of offences against health and safety legislation which have – in differing ways – been subject to processing by regulatory agencies; unlike injury data, offence data require no reconstruction before we can speak of 'crimes'. In this section of the chapter, prosecution and other enforcement data are presented. Since a key concern here is with the quality of these data, and given the sociologically established fact that official data reveal much about the *modus operandi* of the agencies producing that data rather than simply representing measures of some external reality, particular attention is paid to the enforcement practices of HSE.

Prosecution data

Provisional data for the year 1996–97 indicate that 1,256 prosecutions were laid, of which 1,052 (some 84 per cent) were successful. This represents a marked decline in prosecutorial activity, from 1,767 (1,452 successful) in 1995–96, and 2,653 (2,289 successful) in the year 1989–90, the year for which the highest number of health and safety prosecutions are recorded.

Information which is made publicly available does not usually allow any determination of how many of these prosecutions follow occupational *injuries*, though some indications have been given. On the basis of data supplied to him for 1994, Bergman (1994) was able to calculate that less than 40 per cent of workplace deaths from fatal injuries were followed by prosecution, the vast majority of these being brought under sections of the Health and Safety at Work Act which overwhelmingly initiate fines rather than custodial sentences. Data made available for proceedings instituted in 1996–97 and 1997–98 show that 100 prosecutions followed the 696 RIDDOR reportable fatal injuries during 1996–97. This represents a prosecution in about 14 per cent of cases, a marked decline from the figure calculated by Bergman; again, the vast majority proceeded under the HASAW Act (personal written communication, HSE Operations Unit, 2 September 1998).

While these low rates of prosecution may indicate that the vast majority of occupational injuries do not occur as a result of safety offences, they more plausibly reflect the general enforcement strategy of HSE. Thus, low rates of prosecution for safety crimes can be largely explained by both the acceptance of the importance of industry's commercial 'imperatives' by the HSE (Whyte and Tombs, 1998), and its view of itself as a body which must co-operate with and advise industry (Pearce and Tombs, 1998). The effects of the dominance of what have been (misleadingly, Tombs, 1998) called 'compliance strategies' in the enforcement of health and safety legislation are inevitably reproduced in all aspects of HSE activity, so it is no surprise to find relatively low rates of prosecutorial action against safety and health offenders (Sanders, 1985).

A total of 1,256 prosecutions set against almost 200,000 all-reported injuries for 1996–97 is a rather different proportion from 100 prosecutions following 696 deaths in the same year. In other words, even if the rate of prosecution following a fatality remains low, there is a significant disjuncture between the rates of prosecution following fatal as against non-fatal injuries. This difference is not explained by the seriousness of the offence; the overwhelming majority of offences under health and safety law use an inchoate mode (Wells, 1993: 6), that is, they are pursued on the basis of the offence (for example failure to train adequately, or to guard machinery) rather than the outcome (for example a near-miss, minor injury or multiple deaths). That being said, it is unlikely that the outcome of an incident has *no* bearing upon the decision of whether or not to prosecute, be this symbolic or in response to pressure (from politicians, trade unions, work colleagues, relatives, the media and so on).

One crucial explanatory factor can be found in HSE's general investigative activities and procedures following workplace injuries. Two points need to be made regarding investigation.

First, on the frequency of investigation. HSE now investigates all fatalities which occur within a specified workplace – although not all fatalities reportable under RIDDOR 1995 (HSE, n.d.). Fatal injuries are by far the most likely category of injury to be subject to any form of investigation by HSE. By contrast, a survey of enforcement activity in the West Midlands area between 1990 and 1992 found that HSE failed to investigate 79 per cent of all major injuries (Bergman, 1994: 95). With the subsequent introduction of RIDDOR 1995, the vast majority of occupational injuries are reportable by telephone, and HSE will simply not visit the site of the incident to follow up that report. Thus the investigation of non-fatal injuries is now even less likely: in 1996–97, HSE inspectors investigated just *4 per cent* of major injuries (Clement, 1997; Fidderman, 1998: 265).

Second, on the nature of investigations. In general, when the HSE does conduct investigations in the wake of injuries, its primary, often sole, concern is with recommending remedial measures, if it deems any are required, in order to prevent future occurrence of such an event. The HSE certainly does not seek to gather evidence or information upon which any future prosecution might be based. Until recently, this general approach to investigation extended to fatalities also. Thus in his detailed study of 24 deaths at work, Bergman notes, 'There appears to have been no investigation in any of the cases into determining whether there were previous injuries or near misses' (Bergman, 1994: 97). Indeed, deaths at work were distinguished within the criminal justice system as being the only category of 'avoidable' deaths to be exempt from police investigation (Slapper, 1993). Bergman also notes that the HSE and local authority environmental health departments consistently fail to examine the role of senior company officers in relation to deaths at work, since they 'are not viewed as potential criminals whose conduct requires investigation' (Bergman, 1994: 97).

Now, in one important respect, this situation has been changed recently. In spring 1998, a new memorandum of agreement between HSE, the police and the Crown Prosecution Service was published, on the basis of which new guidelines for investigations and prosecution relating to deaths in the work-place are to be implemented (HSE, 1998). While there is little in this agreement which indicates any greater likelihood that criminal-type investigations, and thus prosecutions, will follow, they are at least better facilitated – for the agreement states that 'a police detective of supervisory rank should attend the scene of a work-related death' (HSE et al., 1998: 6). While there are many likely limitations to the effects of this new memorandum (Bergman, 1998), it does represent some improvement on previous practices. For our purposes, though, it really is applicable in so few cases of injury (a small set of fatalities) that prosecutorial data will continue to reveal little about the extent of safety crimes.

It is also impossible to consider the validity of prosecutorial data as an index of safety crimes without noting one of the other key determinants of prosecutorial policy, namely 'the range of sanctions available to the court and current judicial thinking about how the available sentences should be used' (Slapper, 1997: 226; see also Slapper, 1993). Successful cases taken in the higher courts can result in imprisonment or unlimited fines; those taken in magis-trates' courts attract a maximum penalty of £20,000 per conviction. Almost without exception, successful prosecutions for safety offences result in the imposition of a fine. Thus it is of interest that while recent years have seen the emergence of some large fines, and the first uses of custodial sentences, increasing numbers of health and safety offences are now being tried in magistrates' courts. Both Moore (1991) and Bergman (1994) had noted that 90 per cent of all prosecutions for health and safety offences take place in the lower (i.e. magistrates') courts. This figure is now almost 95 per cent (HSC, 1996a: 56, 116). Whilst fines remain the most common formal sanction, both their absolute and relative levels tend, almost without exception, to be derisory, even if these have been subject to recent increase. The average penalty per conviction increased from £410 in 1986–87 to £3,886 in 1997–98 (provisional), and some exceptionally large, high profile fines were imposed during this period.

Other enforcement data

Prosecution is not the only enforcement option available to HSE inspectorates. Offences can also be dealt with by the issuing of various notices as alternatives to prosecution. An Improvement notice requires a breach to be remedied within a specified period of time; a Deferred Prohibition notice is an order that a certain piece of plant or type of work activity will be shut down unless a breach is rectified within a specified period; and a Prohibition notice – the most serious enforcement order – effects an immediate closure of a worksite, part of that worksite or a particular work process until a company complies with legislation of which it is in breach.

The total number of notices issued in any one year does provide a minimal indication of the numbers of safety crimes for that period. We should be wary of investing this data with too much significance, however. While in recent years there has been a steady decline in the issuing of notices, this does not, of course, necessarily reflect any change in the numbers of offences encountered by inspectors; in fact, it says much more about changes in enforcement strategy (and resourcing – see p. 73). For example, in 1991–92, HSE shifted explicitly from the general *modus operandi* of issuing notices for specific breaches, to issuing notices covering several breaches which required changes in, or the introduction of more effective, management, training, and internal compliance procedures; the effect of this is to reduce the total number of notices issued for a fixed number of offences. Further, HSE states that the decline in notices in 1995–96 'reflects the reduction in the number of work-place inspections in favour of an increase in central approaches to firms; these enable inspectors to deal with health, safety and welfare issues at a number of locations by focusing on management systems' (HSC, 1996a: 116). More generally, HSC/E has moved towards a more explicit advisory and educational role in recent years, further abandoning any formal enforcement function (HSC, 1996b: 38; Tombs, 1996).

There are perhaps some very good reasons for the recent marked decline in prosecutorial, investigative and general enforcement activity on the part of the HSE and its inspectorates to be found in the resourcing of HSE. If the HSE has never been granted the resources to act as any kind of police force for industry, the level of resource has recently declined. At 1 April 1976, when the HSE was formally established, there were a total of 3,282 HSC/E staff in post (including temporary staff). This rose to a peak of 4,226 in 1979, before beginning an almost year-on-year decline until the reversal of this trend in 1990, following which a new peak of 4,545 staff was reached in 1994. Since that time, the year-on-year decline in staff has been re-established, so that by 1 April 1999 there were 3,880 HSE staff in total. Moreover, of this total staff, just 1,497 are actually inspectors, the remainder being engaged in a range of support functions. Further aspects of the growing resource pressures on the HSE and its inspectorates are increases in the volume and range of regulations to be enforced, the assumption of new branches of enforcement responsibilities, and the increasing number of smaller, if often short lived, companies.

Several conclusions offer themselves on the basis of this consideration of the use of offence data as a basis from which to measure the extent of safety crimes. First, it is now widely accepted that inspectors from both HSE and environmental health offices neither enter premises in order to seek out violations, nor respond to the vast majority of observed or known offences by resort to formal enforcement action. Except in the case of the most egregious safety offences, enforcement action is invoked only when persuasion, negotiating and bargaining, often over a very protracted period, have proved 'unsuccessful'. Formal enforcement data record only a tiny proportion of an as yet unquantifiable number of safety offences. Second, the ability of inspectorates to act other than in the ways in which they have historically proceeded – using what have been called 'compliance strategies' – is diminishing, first as a result of a series of ideological assaults upon the legitimacy of external regulation of business *per se*, and second, and relatedly, as a result of increased demands on dwindling resources. The best that can be said regarding the extent of safety crimes on the basis of offence data is that the latter provide an absolutely minimal indication of the volume of the former.

Alternative sources of data on safety crimes

Given the limitations of the two data sources encountered thus far, there is a need to consider other sources of data from which some conclusions might be reached about the extent of safety crimes.

Victim surveys

The increasing significance of victimization surveys – both nationally and locally – is well documented. Amongst such surveys, the most well known and important response to the recognition of the enormous problems with official recording of conventional crimes has been to supplement the data provided through 'notifiable offences' with the British Crime Survey. There is no reason why this instrument should not include reference to safety, nor indeed other forms of corporate crimes; however, this is not the case. Thus the BCS, for all its usefulness, reproduces the existing emphasis upon conventional crimes. A somewhat more objectionable point, and a useful indicator of political priorities attached to different forms of offending, is the fact that the Home Office has sponsored a victimization survey on crimes *against* business in England and Wales (Mirrlees-Black and Ross, 1995), while research into such crimes in Scotland is currently being sponsored by the Scottish Office. Further, it will be very surprising if more than a tiny handful of the local audits required under the Crime and Disorder Act 1998 include any reference to corporate crimes in general, or safety crimes in particular.

More generally, there have been barely any efforts to generate more accurate safety crime data via victim-based reporting, and there are some valid reasons for this, not least the very different relationships between offender and victim that tend to characterize corporate crimes in general when compared with conventional crime (see Tombs, 1999). However, there have been some attempts to measure such victimization. *The Second Islington Crime Survey* included questions relating to commercial crime, to pollution and to health and safety offences – specifically, with regard to the latter, asking whether respondents had sustained (non-fatal) injuries whilst in paid employment during the previous 12-month period. One limitation of this data is that only a sub-set – 889 – of those questioned in the survey were asked questions regarding occupational injury, while just 58 per cent of these had been in paid employment in the previous 12 months. Given the small number of respondents, and the fact that many of these worked in relatively high risk occupations (textiles and construction), it is difficult to make any generalizations based upon the responses. However, 5 per cent – 25 respondents – stated that they had been injured at work in the previous 12 months so that, based upon the then current reporting requirements (RIDDOR '85), these figures produced an injury rate some 30 times the national average (Pearce, 1990: 19–20). More generally, the survey as a whole did produce evidence of widespread victimization (Pearce, 1990).

Of greater interest are data produced by recent Labour Force Surveys. The Labour Force Survey (LFS) is a survey of some 60,000 private households in Britain, based upon a systematic random survey design, and carried out by the Office for National Statistics. Since 1992 it has been conducted on a quarterly basis, and HSE has asked four questions on (non-fatal) workplace injuries at each winter-quarter survey since 1993–94; HSE data are available for 1989–90, and then for each year from 1993–94. At the most general level, 1995–96 LFS data indicate that 1.060 million people suffered a workplace injury, 403,000 of which would have been reportable to HSE under RIDDOR '85; the latter figure contrasts with the 164,288 injuries recorded under RIDDOR for that same year (HSC, 1997: 137). It is on this basis that HSE is now able to produce estimates of the extent of under-reporting within industries, thus proving an important adjunct to data collected under RIDDOR (see Stevens, 1992; HSC, 1997: 68–79).

The Labour Force Survey data are of interest in their own right, since they indicate a widespread failure on the part of employers to meet the legal duty to report incidents, an offence under safety law. Yet while the LFS is a significant adjunct to, and improvement upon, data collected under RIDDOR in terms of measurement of injuries, it still does not reveal much of use about the levels of *crimes* inherent in these injuries.

Self-report surveys

A further option for uncovering the hidden figure of crime is the use of self-report surveys. The limitations of such surveys are well known, and there is even less reason to be confident in the likelihood of employers or corporate offenders responding accurately to self-report-style questions than the young people upon whom such techniques are most frequently used.

One reason for this pessimism is that employers significantly under-report at present within RIDDOR, a reporting scheme which in many respects is a self-reporting system, if not strictly analogous to self-report surveys. If employers fail to report where there is a legal obligation to do so, and despite efforts by HSE to make such reporting simpler and less time-consuming, then there is room for scepticism regarding the utility of self-report surveys here.

A second problem with self-report surveys in the context of safety crimes is the existence of particular pressures upon employers which are likely to prevent them revealing the hidden figure of unreported injuries, even in the context of reassurances of anonymity. Injuries – and, more generally, accidents – are important for employers, as they are often linked to the levels of premiums payable for various forms of insurance, are one basis upon which HSE develops inspection priorities, and are often linked to inter-organizational bonus or penalty schemes (written, for example, into contracts between main and subcontractors). Each of these points help to explain why there is documented evidence of employers misleading outside agencies regarding the actual levels of injuries. Further, the pressures emanating from the incentives to present low injury rates can also be transmitted to employees – thus it is known that award or bonus schemes for achieving a certain number of employee hours without a lost-time accident create peer group pressures to conceal injuries from both management and external agencies.

Whistle-blowers

If there is either direct or more subtle manipulation of injury data by employers, access to internal whistle-blowers may be one means of checking, and developing more accurate, data. The likelihood of whistle-blowers being prepared to risk their employment status is a relatively slim one, of course, whilst once 'on the outside', the access that whistle-blowers can offer, and indeed the legitimacy which might be extended to their information, is diminished. In general, the use of whistle-blowers raises problems about the receptivity of powerful audiences to arguments based upon information thus supplied. However, there is no doubt that such individuals can perform a useful role in establishing more accurate understanding of levels of injury, if not crimes. Of course, data offered by whistle-blowers is likely to be of relevance on a company-, or at best an industry-wide, basis; resort to individuals will do little to develop more reliable, utilizable data on either injuries or safety crimes across sectors.

Both Whyte, and Woolfson and colleagues, have been engaged in exemplary work in the UK offshore oil industry, attempting to uncover hidden incidents of lost-time accidents in particular, and indeed evidence of safety offences offshore more generally. Each has used a range of methods to gain access to individuals wishing to report instances of safety violations or injuries which have neither been reported to, nor become known of by, HSE. Whyte has relied particularly upon extensive interviews with offshore workers, conducting these onshore, away from places of work, in order to encourage discussion of incidents and injuries that should have been reported (Whyte, 1998, 1999). Woolfson and colleagues have developed a mutually productive working relationship with the independent offshore oil workers' union the OILC, itself formed largely to progress issues around occupational safety (Beck et al., 1998; Woolfson et al., 1996).

Two consequences follow from these researchers' efforts, and their use of whistle-blowers. First, each area of research has met with resistance from oil companies, of course, but also from HSE. Woolfson and his colleagues have found themselves vilified by the oil operators' organization UKOOA; Woolfson has also appeared on an HSE 'greylist', an internal memorandum requiring that his requests for information to that government body be monitored (Monbiot, 1998; Woolf, 1998). Whyte has also been subject to HSE pressures; he has recorded an account of a senior HSE employee making less than subtle reference to a large amount (£6 million) of HSE research funds which might find its way to his home university were certain conclusions not to be reached in any research thus funded (Gibb, 1998; Whyte, 1999).

Second, more particularly, both researchers have received criticism for their methodologies – in short, for their reliance either upon whistle-blowers (politically motivated individuals) or the OILC (a politically motivated workers' organization) – bearing testimony to Nelken's warning that reliance upon whistle-blowers and investigative journalists opens academic work to criticisms of 'accuracy, frequency, or representativeness' (Nelken, 1997: 892). Whyte has been criticized for his use of interviews with offshore workers on trains or in bars. Despite this being the most sensible – perhaps the only – method of making contact with a mobile, relatively transient, and often highly intimidated workforce, whose workplace is, quite literally, cut off from 'outsiders', the use of such sampling techniques and locations has opened up his work to charges that it is less than rigorous, a charge based upon erroneous, but still dominant, assumptions regarding what constitutes 'scientificity' in social research. And, it is much easier to make this charge stick when the conclusions of the work at which it is aimed are highly critical of relatively powerful organizations – here, oil companies, regulators, and indeed some trade unions.

Conclusion

Attempts to measure the scale of safety crimes are fraught with problems. Many of these are similar problems to those encountered by researchers aiming to uncover the so-called dark figure of hidden, conventional crime. Yet, almost always, those considering safety crimes have an extra set of obstacles to negotiate before engaging in quantitative talk.

It is possible to point to ways in which existent injury data may be reconstructed. This reconstruction is perhaps analogous to the attempts to uncover unrecorded crimes. Yet if one wishes to speak of – and measure – safety *crimes*, even reconstructed and fuller injury data form only raw material on the basis of which measures of crimes might be developed. Injuries are not crimes, and in using the former to represent the latter the enterprise of speaking of safety crimes becomes problematic. For this exercise is conducted in a harsh social and political environment, and the act of manipulating data to reach – albeit informed – estimates of safety crimes opens up researchers to charges of bias, moral entrepreneurship and so on (see Slapper and Tombs, 1999). The considerable, but necessary, reconstruction of data is precisely what makes the results of this process vulnerable to such charges. By contrast, the use of enforcement data would seriously understate the extent of safety crimes, and would not allow a researcher to speak of the consequences of such crimes – thus allowing the charge that many offences on which enforcement action is taken are minor, technical breaches of regulations, rather than phenomena accurately captured by that morally charged label 'crime' (Pearce and Tombs, 1998; Slapper and Tombs, 1999). Further, simply to use enforcement data is to remain at the level of counting the effects of a *modus operandi* of enforcement agents, a process long rejected by almost all critical social scientists.

In the light of these difficulties, then, what is to be done? Two obvious solutions present themselves. The first is to abandon attempts to generate utilizable quantitative estimates of the scale of safety crimes, and to engage only in qualitative work. There is now a growing body of literature around occupational safety, not least focused upon and emanating from Britain, and upon which criminologists may draw (even if much of this does not emanate from criminology). From this body of work it is possible to discern much about the normality and routineness, the process, the causes, the unequal distribution of effects, and the appropriate modes of regulating and 'punishing' such offences (see Slapper and Tombs, 1999). Some of this work focuses upon particular forms of offence (notably corporate manslaughter), on particular industries, upon particular types of incidents, or even one-off 'accidents'. This body of work is an important resource. However, without descending into the dualism (and false standards of scientificity) of advocating qualitative work only when quantitative efforts become overly problematic (Bryman, 1997), there is a real problem with a complete embrace of the qualitative. For to abandon the terrain of quantification is to vacate a key ideological space within which social problems in general, and crime in particular, are perhaps most subject to debate. Moreover, it is to relinquish somewhat the potential leverage over policy-makers which resort to quantification can often facilitate.

The second possible solution is to abandon the consideration of safety crimes at all, that is, to vacate the terrain of crime and criminology altogether. This would allow *some* quantification to proceed – for example the production and use of estimates of the numbers of occupational injuries, major injuries and deaths – without the researcher having then to engage in the technically and 'morally' fraught issue of converting such data into data on crimes, risking the eventual undermining of the enterprise. At a pragmatic level, this has some attraction. Certainly one of the problems – it is an irritation, of course, but it is more than this – in presenting any argument (empirical, conceptual or theoretical) about safety crime within the discipline of criminology is the almost ever-present prior need to justify the very object of the study, that is, to attempt to establish that phenomena such as 'safety crimes' actually exist; at present, the phrase 'safety crimes' has only a tenuous foothold on the terrain marked out by 'criminology'. This attraction notwithstanding, such a strategy has real, and unwelcome, consequences. Perhaps most significant of all is the point at which this chapter began, which is the realization that contemporary crime debates favour argument based upon quantitative data. In a political sense, one of the points of thinking, researching, writing about, and attempting to measure safety crimes is to point to an area of existent law subject to routine violation which attracts little political, popular or indeed academic attention, and at the same time to throw some light on the activities of the powerful and draw attention to the fact that persistent images of 'the criminal' and 'crime' are both inaccurate and mystificatory. To discard the referent object of crime is to abandon these political efforts. I remain reluctant to do so, since it seems that such an abandonment entails abandoning one means of expressing and furthering a commitment to equity and social justice in public policy on crime.

Suggested readings

Bergman, D. (1994) *The Perfect Crime? How Companies Can Get Away with Manslaughter in the Workplace*. Birmingham: West Midlands Health and Safety Advice Centre.

Pearce, F. and Tombs, S. (1992) 'Realism and corporate crime', in R. Matthews and J. Young (eds) *Isssues in Realist Criminology*. London: Sage. pp. 70–101.

Slapper, G. and Tombs, S. (1999) *Corporate Crime*. London: Longman.

Wells, C. (1993) *Corporations and Criminal Responsibility*. Oxford: Clarendon Press.

Whyte, D. (1999) 'The politics of offshore safety research', in R. King and E. Wincup (eds) *Handbook of Criminology and Criminal Justice Research*. Oxford: Oxford University Press.

Notes

1. Local authorities also have some enforcement responsibilities under health and safety legislation; injury data supplied by HSE include those in local authority enforced

sectors (and see HSC, 1998). Studies of their *modus operandi* have noted that these agencies, like HSE, also adopt primarily a compliance-seeking enforcement strategy (Hutter, 1988, 1997).

2. Thanks to Gary Slapper for first bringing these points to my attention.

References

Beck, M., Foster, J., Ryggvik, H. and Woolfson, C. (1998) *Piper Alpha Ten Years After. Safety and Industrial Relations in the British and Norwegian Offshore Oil Industry*. Oslo: Centre for Technology and Culture, University of Oslo.

Bergman, D. (1994) *The Perfect Crime? How Companies Can Get Away with Manslaughter in the Workplace*. Birmingham: West Midlands Health and Safety Advice Centre.

Bergman, D. (1998) 'Bosses get away with murder', *New Statesman*, 6 November: 29–30.

Bibbings, R. (1996) *Managing Occupational Road Risk: Discussion Paper*. Birmingham: RoSPA.

Box, S. (1983) *Power, Crime and Mystification*. London: Tavistock.

Bryman, A. (1997) 'Quantitative and qualitative research strategies', in T. May and M. Williams (eds) *Knowing the Social World*. Buckingham: Open University Press. pp. 138–156.

Clement, B. (1997) 'Work injuries not being investigated', *The Independent*, 13 November: 15–16.

Fidderman, H. (1998) 'The HSB interview. Mood music with Jenny Bacon', *Health and Safety Bulletin*, 265, January: 10–15.

Gibb, B. (1998) 'UK attacked on safety funding. Researcher claims cash is held from critical studies', *Press and Journal*, 27 July.

Health and Safety Commission (1996a) *Annual Report 1995/96*. Sudbury: HSE Books.

Health and Safety Commission (1996b) *Health and Safety Statistics, 1995/96*. Sudbury: HSE Books.

Health and Safety Commission (1997) *Health and Safety Statistics, 1996–97*. Sudbury: HSE Books.

Health and Safety Commission (1998) *National Picture of Health and Safety in Local Authority Enforced Industries 1998*, London: HSC/Government Statistical Service. Sudbury: HSE Books.

Health and Safety Executive (1996) *A Guide to the Reporting of Injuries, Diseases and Dangerous Occurrences Regulations 1995*. Sudbury: HSE Books.

Health and Safety Executive (1997) *Press Release E133.97 – 28 July, 1997. Headline Workplace Health and Safety Statistics 1996–97*. Sudbury: HSE Books.

Health and Safety Executive (1998) *Press Release E87:98 23 April 1998. Work-related Deaths – a Protocol for Liaison*. Sudbury: HSE Books.

Health and Safety Executive (n.d.) *Accident Investigation Practice*, supplied by personal communication, HSE Policy Unit, 25 August 1998. Sudbury: HSE Books.

Health and Safety Executive, Association of Chief Police Officers, Crown Prosecution Service (1998) *Work-Related Deaths. A Protocol for Liaison*. Sudbury: HSE Books.

Hutter, B. (1988) *The Reasonable Arm of the Law? The Law Enforcement Procedures of Environmental Health Officers*. Oxford: Clarendon Press.

Hutter, B. (1997) *Compliance: Regulation and the Environment*. Oxford: Clarendon Press.

Levi, M. (1997) 'Violent crime', in M. Maguire et al. (eds) *The Oxford Handbook of Criminology*, 2nd edition. Oxford: Clarendon Press. pp. 841–889.

Maguire, M. (1997) 'Crime statistics, patterns, and trends: changing perceptions and their implications', in M. Maguire et al. (eds) *The Oxford Handbook of Criminology*, 2nd edition. Oxford: Clarendon Press. pp. 135–188.

Mirrlees-Black, C. and Ross, A. (1995) *Crime against Retail and Manufacturing Premises: Findings from the 1994 Commercial Victimisation Survey*. Home Office Research Study No. 146. London: Home Office Research and Statistics Directorate.

Monbiot, G. (1998) 'Unsafe to criticise', *The Guardian*, 21 May.

Moore, R. (1991) *The Price of Safety*. London: Institute of Employment Rights.

Nelken, D. (1997) 'White collar crime', in M. Maguire et al. (eds) *The Oxford Handbook of Criminology*, 2nd edition. Oxford: Clarendon Press. pp. 891–924.

Nichols, T. (1989) 'The business cycle and industrial injuries in British manufacturing over a quarter of a century: continuities in industrial injury research', *Sociological Review*, 37 (3): 538–550.

Nichols, T. (1994) 'Problems in monitoring the safety performance of British manufacturing at the end of the twentieth century', *Sociological Review*, 37 (3): 538–550.

Nichols, T. (1997) *The Sociology of Industrial Injury*. London: Mansell.

Pearce, F. (1990) *Second Islington Crime Survey: Commercial and Conventional Crime in Islington*. Middlesex: Centre for Criminology, Middlesex Polytechnic.

Pearce, F. and Tombs, S. (1992) 'Realism and corporate crime', in R. Matthews and J. Young (eds) *Issues in Realist Criminology*. London: Sage. pp. 70–101.

Pearce, F. and Tombs, S. (1998) *Toxic Capitalism: Corporate Crime and the Chemical Industry*. Aldershot: Ashgate.

Reiman, J.H. (1979) *The Rich Get Richer and the Poor Get Prison: Ideology, Class, and Criminal Justice*. New York: John Wiley and Sons.

Robens, Lord (1972) *Safety and Health at Work. Report of the Committee 1970–1972*. London: HMSO.

Royal Society for the Prevention of Accidents (1998) *Managing Occupational Road Risk*. Birmingham: RoSPA.

Sanders, A. (1985) 'Class bias in prosecutions', *The Howard Journal*, 24 (3), August: 176–199.

Slapper, G. (1993) 'Corporate manslaughter: an examination of the determinants of prosecutorial policy', *Social and Legal Studies*, 2 (4): 423–443.

Slapper, G. (1997) 'Litigation and corporate crime', *Journal of Personal Injury Litigation*, 4: 220–233.

Slapper, G. and Tombs, S. (1999) *Corporate Crime*. London: Longman.

Start, R.D., Delargy-Aziz, Y., Dorries, C.P., Silcocks, P.B. and Cotton, D.W.K. (1993) 'Clinicians and the coronial system: ability of clinicians to recognise reportable deaths', *British Medical Journal*, 306, 17 April: 1038–1041.

Start, R.D., Usherwood, T.P., Carter, N., Dorries, C.P. and Cotton, D.W.K. (1995) 'General practitioners' knowledge of when to refer deaths to a coroner', *British Journal of General Practice*, 45: 191–193.

Stevens, G. (1992) 'Workplace injury: a view from HSE's trailer to the 1990 Labour Force Survey', *Employment Gazette*, December: 621–638.

Tombs, S. (1996) 'Injury, death and the deregulation fetish: the politics of occupational safety regulation in UK manufacturing', *International Journal of Health Services*, 26 (2): 327–347.

Tombs, S. (1998) 'Tony Prosser: law and the regulators; Julie Black: rules and regulators', *Journal of Law and Society*, 25 (3): 452–460.

Tombs, S. (1999) 'Health and safety crimes and the problems of knowing', in P. Davies, P. Francis, and V. Jupp (eds) *Invisible Crimes: Their Victims and Their Regulation*. London: Macmillan. pp. 77–104.

Tombs, S. (1999a) 'Death and work in Britain', *Sociological Review*, 47 (2): 345–367.

Wells, C. (1993) *Corporations and Criminal Responsibility*. Oxford: Clarendon Press.

Wells, C. (1995) *Negotiating Tragedy: Law and Disasters*. London: Sweet and Maxwell.

Whyte, D. (1998) 'Overcoming the fear factor', *Public Money and Management*, October–December.

Whyte, D. (1999) 'The politics of offshore safety research', in R. King and E. Wincup (eds) *Handbook of Criminology and Criminal Justice Research*. Oxford: Oxford University Press.

Whyte, D. and Tombs, S. (1998) 'Capital fights back: risk, regulation and profit in the UK offshore oil industry', *Studies in Political Economy*, 57: 73–101.

Woolf, M. (1998) 'Civil servants are told to turn spy', *Observer*, 3 May.

Woolfson, C., Foster, J. and Beck, M. (1996) *Paying for the Piper: Capital and Labour in Britain's Offshore Oil Industry*. London: Mansell.

4

DOING INTERVIEWS WITH FEMALE OFFENDERS

Pamela Davies

Contents

This chapter is about planning and doing research through interviewing. In particular it draws on experience of conducting interviews with female offenders in the north east of England. Interviews range from the informal, unstructured discussion through to very structured formats with answers offered from a prescribed list in a questionnaire or interview schedule. At one extreme interviews can be conducted as conversations and at the other extreme they involve little interaction between the researcher and the researched. An example of the latter is Computer Assisted Personal Interviewing (CAPI) where interviewers enter responses into a lap-top computer or respondents use the computer to answer questions themselves – this is known as self-keying by respondents (Mayhew, 1996). The research described in this chapter moves beyond 'conversation' but remains far removed from the very structured end of the interview scale. This research chooses to adopt a semi-structured interview format. Such a format lends itself to an appreciative stance consistent with a feminist approach. For a fuller description of these methods see Jupp (1989), Mason (1996) and Sapsford and Jupp (1996).

This chapter focuses upon the contextual factors surrounding the whole experience of doing qualitative interviews. These elements have not traditionally formed a substantial part of the text in accounts of research or in papers which report findings. Where accounts of this nature have been provided by authors and researchers they are tucked away in a short appendix or a 'note on sources' at the end of a book or they are highly generalized (McCracken, 1988). In more recent mainstream publications such accounts of research experiences have become more prominent and transparent. Acknowledging these aspects of doing research is routine practice for feminist-inspired work particularly that on women and crime in both America (Daly, 1994; Maher, 1997) and Britain (Carlen, 1988; Carlen et al., 1985; Hudson, 1994). Over and above this there are also some wide-ranging and challenging debates surrounding feminist perspectives in criminology (Gelsthorpe and Morris, 1994; Naffine, 1997). This body of work is concerned with renewing theoretical debates about the feminist critique, with feminist methodologies (Gelsthorpe, 1994) and with philosophy and epistemologies (Cain, 1994). Such debates are beginning to reveal more about the various 'truths' of doing research and also the value of doing less research and more analysis informed by theory.

In official publications such as those emanating from the Home Office and police sources we are presented with an uncluttered, clean and sanitized version of the research findings clearly set out for us to 'read off'. Rarely does the account ask us to think about what goes on behind the scenes and about the often painful and slow process of the research experience – how the conclusions are 'arrived at' and what the stories are along the way. Examples of researchers and practitioners 'telling it like it is' include exposing the place of subjectivity in fieldwork-based research. Research is often not a neat and tidy process and the day to day business of conducting it can be frustrating if it is not accepted that there are fuzzy edges to such work. The most common way of learning about these aspects of research is orally, through talking about experiences of fieldwork with fellow colleagues and students.

Such accounts of experiences give us added value in the form of observational and 'reflective' data in addition to the 'literal dialogue' (Mason, 1996) as well as encouragement and reassurance as researchers and students. More traditional and common 'methods' and 'good research practice' instruction comes in textbook format (Marsh, 1988; Oppenheim, 1992) and is littered with warnings about how not to do research. After reading such texts students and prospective researchers may be easily discouraged from embarking upon any form of research. Awareness of the guiding principles of doing good research is essential but some readers ought, and may prefer, to be enticed into doing research by reading about how it is actually done.

The chapter will briefly describe the context of the research to be discussed. It will then go on to discuss how various research methods and tools were employed. In particular the planning and conduct of research as a continuous process of decision-making is highlighted.

The context: women and crime for economic gain

This chapter relates to a part of the research that has investigated women who commit crime for economic gain (see Davies, 1999). Whilst the field-work has encompassed various different strategies and stages of gathering data, the major part of it has involved one-to-one interviews with women who have committed 'economic crimes'. Where specific terms are used in the formulation of research problems it is important to define these in the context of the project. In this case the definition of 'economic crimes' includes most varieties of property crime such as thefts (including shoplifting and pilfering, frauds and forgery, as well as many crimes that come under the general heading of white-collar crimes), burglary and car crimes as well as prostitution and drug-related offences. These crimes are more likely to contribute to the illegal marketplace and have been referred to as 'crime for gain' (Field, 1990).

Interviews were conducted with 26 women in total. Twenty-one interviews took place in prison and five in the community with women who were either on probation or were no longer under the jurisdiction of the courts. All the women interviewed were white and can be loosely described as working class. The majority were in their twenties although their ages ranged from 17 years to 46 years. Sixteen of the women had a total of 33 children between them with two of these same women pregnant again in prison. Thirteen had a partner whilst 13 described themselves as single.

All but one of the women was serving a sentence of the court or a court order and all had been convicted of at least one criminal offence, and in the majority of cases several. Dispositions ranged from being remanded in custody and awaiting trial to prison sentences ranging from three months to two years and probation supervision with and without various conditions.

The crimes the women were most frequently engaged in, and discussed in interviews, were thefts, in particular shoplifting, fraud and deception but also employee and car theft. Burglary, drugs and prostitution-related offences also figured significantly in their offending profiles whilst a variety of other offences were also included in the range of crimes they had committed. Summarizing the female offenders interviewed, these women would generally be characterized as 'hustlers' (Campbell, 1991; Maher, 1997), economically marginal and committing petty offences (Steffensmeier and Allen, 1996). Crime appeared to constitute a major source of income for these women.

Before the interviews

Before interviewing takes place much research, organization and planning are needed. Indeed by this stage many important decisions about the research have already been made. For example decisions about whether to conduct personal interviews or administer a postal survey or other less grounded methods of research. Decisions then need to be made about how to introduce the research and how best to describe what it is all about to the various gatekeepers as well as the interviewees. Decisions need to be made about how to organize the interviews including which gatekeepers to approach for access and about where interviews might take place. It is important to consider how the interview will be conducted. Time and length restrictions as well as other non scheduled interruptions may affect the way in which the interview is conducted and strategies need to be thought through in order to achieve a good interview. Additionally decisions need to be made regarding the nature of the qualitative interview itself including whether or not tape recording will be an option or whether copious note taking will be required. These decisions regarding research, organization and planning before interviewing require careful thinking about and negotiation with others before the most appropriate avenues are prioritized, or more likely compromised upon, by the researcher. Such decisions are part of the forward planning process (see Francis in Chapter 2) and many of them will be altered, modified or even abandoned when the interview becomes a reality. Flexibility needs to be built into the early preparations.

In the period before my own interviewing took place several decisions were made regarding the type of interview to be conducted. The decision to employ personal interviews and a semi-structured interview format was largely based upon the nature of the research questions. These questions included exploring the types of crimes the interviewees engaged in and details of the ways in which they conducted them. Pilot discussions had shown similarities between the women in respect of the types of crimes committed. There were also commonalities between the ways in which the crimes were executed. These factors could have suggested a structured format but a further decision made early in the research process concerned a preference for keeping the research grounded and for maintaining respect for the narratives of the individual women as far as possible. Grounded research might involve several of the following principles: collecting data at first hand from informants; collecting it in their own terms and in the light of what they think is significant (not what pre-existing theory thinks is significant); and developing theory grounded in the actor's words and/or actions (Glaser and Strauss, 1967). I hoped the research would elicit common data but also differences and variations whilst remaining faithful to the sources from which the information came. These latter points also raise issues pertinent to doing feminist research.

Feminist research

Feminist research practice has been explored by various writers (Gelsthorpe and Morris, 1994; Maynard and Purvis, 1994; Naffine, 1997; Stanley and Wise, 1993) and many others have brought their own feminist perspective to bear on their fieldwork and analysis of different areas of sociology (Carlen, 1988; Carlen et al., 1985; Heidensohn, 1985; Maher, 1997; Walklate, 1996). This work demonstrates that there is no single feminist viewpoint or perspective. Nevertheless as Naffine has argued:

> Many feminists are of the view that the angle from which the dominant class views the world, is one which provides a poor field of vision. Subjugation, and reflection upon that status, makes for a better appreciation of the world. (Naffine, 1997: 51)

This particular view is similar to the approach to research that is grounded. Decisions about doing qualitative interviews with female offenders were influenced by the attempt to follow good reflective research practice – which some call feminist research – but other issues add to its feminist orientation. These issues include conscious choices made at the outset by a female researcher; the fact that all the informants were female; and that the subject matter explores a research question that has a specific gender bias. The research also employs the use of the semi-structured in-depth interview method that is generally seen as consistent with feminist research because such interviews seek not to be exploitative but to be appreciative of the position of women.

Access, safety and politics

Further into the research process more issues present themselves, such as gaining access to interviewees as well as safety concerns for the interviewer. Politics may complicate the research process as well as practical difficulties such as finding suitable interview locations.

Initial access to prison-based interviewees was made surprisingly easy by the co-operation of staff members at a regional remand centre. Prisons have traditionally been 'closed' institutions (see Martin in Chapter 12) so gaining access for research purposes has been troublesome (Cohen and Taylor, 1972) and continues to be a considerable undertaking for most. Research efforts can often be thwarted once carefully planned research is due to be operationalized. Confidence and encouragement follow from early successes in fieldwork – such as gaining access to women in prison. Further access issues are discussed later.

The period between the preparation and the actual doing of the interviews can be a time of anticipation, which may include moments not only of concern about what lies ahead, but also of fear. Although fieldwork is often a part of the research process to look forward to, doing interviews with known criminals who have apparently done something serious enough to warrant their loss of liberty can still be a daunting prospect. Some such feelings and concerns come under the general title of safety issues. In respect of who to interview it is possible to exercise some measure of control over the selection of prisoners (see also Martin in Chapter 12). In respect of the type of offence allegedly committed by the offender, the offence categories which were of interest were made explicit. It was emphasized that economic crimes only – that is fraud and forgery, thefts and shoplifting-type offences (hence excluding violent women and violent offences including murder) – were to be investigated. Such parameters can be built into the research design (see Francis in Chapter 2) for genuine reasons related to the research question but they also help to assuage concerns about safety at the pre-interview point. Another concern when conducting prison-based research is about being taken hostage, particularly if there are ongoing grievances in particular establishments or current political issues relating to the treatment of prisoners. Some of these safety issues will be discussed further in the next section.

Another difficult decision concerns the offering of inducements to interviewees. Should payment be offered to the women for the interviews? Others conducting research with impoverished women have made payments and there are valid arguments both for and against this practice. For many this is really not an option, as funding is often not available but other 'inducements' might be considered. In carrying out the research small quantities of cigarettes were taken to all interview venues together with a lighter (with permission from wing staff in the prison). This practice could be seen as part of the 'research bargain' and proved particularly useful in the prison setting.

Lesser concerns were related to a fear of the unknown: what to wear? What to take with me? Whether I would arrive on time. Whether the prison gate and security had been informed of my arrival and my business. Whether there would be any obstacles to going in and getting on with the interviews. How I would manage to conduct myself appropriately and credibly with both staff and inmates. Although many worries and concerns are common to both prison- and community-based interviews, some matters are location specific. For example, although gaining access to women in the institutional setting was not entirely free of difficulties, access to women who had offended, or who continued to offend and who (still) had their freedom and liberty, can prove a slower and more cumbersome process of negotiation. In this experience of fieldwork planning inroads were made at the most senior levels of the probation service, as well as the most senior levels of the remand centre. In principle, access was granted by both prison and probation service managers. However, moving from one level of access to another within the probation service proved time consuming and much patience and persistence were required to secure the interviews.

Several issues compounded the delay in gaining access to interviewees in the community via probation networks. Access had been approved and suggestions made as to how to progress the fieldwork by members at senior management levels including the Chief of Probation and the Information Officer. Meanwhile contacts were also made and co-operation enlisted from sources at ground level in the service, from probation officers who experienced routine and direct contact with female offenders on their caseload. However, it was found that support and clearance were also required from middle management, namely team leaders in local areas. This layer of personnel also needed to be 'put in the picture' and to be kept abreast of what was going on and given the opportunity to give approval. As a researcher eager and impatient to follow up early successes and maintain the momentum of the interview process, getting the community-based probation introductions proved hard work. At times it was felt that 'they' (middle management) were hoping I would give up and go away. Phone calls were often frustrating and made in vain or resulted only in my being asked once more to provide written details of the research in general and what was required of the probation service in particular. Follow-up telephone calls often seemed to 'pass the buck' on to someone else, or the individual it was necessary to speak with was on holiday or sick leave. The reality of doing grounded research often means revising carefully devised timetables and the constant need to renegotiate highlights the importance of being flexible and spontaneous.

In my own experience and with hindsight it was clear that there were other complicating factors at work during this period of time in the probation service. A review process was causing tremendous uncertainty and insecurity in the service generally and amongst individual employees about their jobs and futures. This unsettling situation was having its effects upon one part of the service after another throughout the fieldwork period when morale was low and long-term sickness high. Nevertheless, despite the difficulties caused by outside influences and the politics within the probation service, appointments were successfully secured to visit probation-run women's groups as a prelude to securing one-to-one interviews via snowball sampling.

Setting up interviews

Snowball sampling is a way of selecting a sample that is akin to a 'chain letter'. In this experience of fieldwork initial contact with one or two willing interviewees was achieved and these women put me in touch with other women also willing to be interviewed. In this way snowball sampling provides a self-selected sample. It is also an acceptable and ethical way of sampling, although there are problems associated with this method in respect of typicality, representativeness and bias (see Mason, 1996 for more on sampling and selection considerations generally).

Having identified several female (ex) offenders in the community willing to be interviewed, further thought and planning were required prior to and during the interview process. For example, if probation-building space was not available or appropriate, alternative venues would be required in which to conduct the interviews. At this stage practical, ethical and safety concerns present themselves. Several options were considered. I briefly entertained, but soon dismissed, the idea of bringing women to my own office at the university. Although a private and safe place from my point of view it was felt that it would be an alien and strange setting for the interviewee. Also, I could not guarantee their privacy because there were colleagues and students around – many of whom knew I was in the throes of doing interviews in connection with my research interests. Ethical considerations of anonymity and confidentiality would be severely compromised.

Another venue which the women may have viewed as the interviewer's 'turf' – that of my own home – was intermittently considered in my desperation of finding anywhere to interview some of the women. The risks of imparting my name and home address – or any other personal details – to known offenders, and inviting them into my own home were keenly felt. However, this would have been a simple solution to my problem and might have accrued a number of beneficial side effects in respect of gaining access and a good relationship with the women. In the event a more neutral and safe venue was considered appropriate and I was never tempted to resort to this option.

Two other options remained real possibilities, both of which could be considered more neutral territory. First, a local café. This option was perhaps the closest to a compromise in terms of 'home versus away' although the choice of café would be an important decision as privacy could be severely compromised even if safety would not be put at risk. The second option might be the interviewee's home. If the opportunity presented itself their own homes might be visited despite some safety concerns. Decisions about where these interviews would take place, as with many of the other issues discussed above, could not all be resolved until the period very soon or immediately prior to the interview itself. Experience illustrates the need to be flexible and to have different options and alternative courses of action in mind as back-up plans.

During and after the interviews

During the interview process issues concerning safety, inducements, sampling, validity and reflexivity arise.

Safety revisited

Some safety concerns can be assuaged by being assiduous about informing home and work of your whereabouts and likely time of return. In this experience the majority of the interviews took place in prison, several took place on probation premises, some in a café and one woman was interviewed several times at length in her own home. Each interview location brought with it its own particular idiosyncrasies. Interviews in prison dominated my early experiences of interviewing female offenders and some further safety considerations were addressed. Several prison staff members were aware of my presence on a wing and either a member of staff was always within earshot or a panic button was available nearby.

Sampling revisited

Having described to the officers on the wing the types of offence (and therefore I hoped the type of offender) I was researching, the selection of inmates to approach was initially at their discretion. In the event the vast majority of the women had at some time committed offences relevant to the research. This was mostly established early on in the interview as the purpose of the discussion was described, and it was soon discovered what each of the women was currently being detained for. On the first afternoon of interviewing, however, I was introduced to a diminutive woman who was on remand for alleged murder. We talked about her circumstances and her life in general before the meeting ended, whereupon she politely thanked me for spending time talking with her. The member of staff as usual after each interview was curious to know how it had progressed. *He* was clearly amused and was testing out my reactions and how I would cope. It was all done in good humour and although the interview had proved an interesting experience it was not useful for the purposes of gathering hard data. After this occasion greater control was exercised over the selection of interviewees in prison. This was achieved by asking women who were particularly helpful and forth-coming about their committing of 'economic crimes' to suggest other inmates' names to me for interviewing purposes. This snowball sampling method was a more grounded form of research practice, but a degree of collusion and connection with the inmates was also achieved that allowed them to become self-selecting. Several inducements clearly contributed to this pattern of recruitment.

Inducements revisited

In the prison setting inducements come in many forms, especially for those women sentenced rather than on remand. The prison grapevine works quickly and efficiently. Not only was there a 'Miss' (later I became Pam) who wanted to talk to them – which held out the prospect of getting them out of cleaning the floor, or their cell, or simply doing nothing – but she had cigarettes with her. This spread of rumour, interest and curiosity all combined to my advantage, as I was becoming known in the prison. Self-selecting interviewees were presenting themselves whilst checks and balances within the interview schedule were ensuring that the stories were their own and individual accounts were emerging.

Validity and reflexivity

This last point raises the question of validity. Good research is valid research. Validity is 'the design of research to provide credible conclusions' (Sapsford and Jupp, 1996: 1). To be concerned about validity is also part of the process of reflexivity. This is important in qualitative research (Hammersley and Atkinson, 1983). In this fieldwork, validity checks on some of the data obtained could have been achieved using a combination of different sources such as both interview data and offender records. Combining these forms of data is a form of triangulation. This, together with supporting literature and a reflective practice generally, would enable very robust checks on the reliability and validity of the research. Such lengths were not deemed necessary in this fieldwork as the data of interest concerned how they carried out their crimes including precise methods of operating, choices of premises and targets and offender networks information, which is not reliably and consistently collated and stored by official sources on female offenders.

The interview process

Once in the one-to-one interview, interviewing skills are important. The interviewer has to do, say and think about several things all at once. Most important is to listen, then prompt and encourage when appropriate but without 'leading', and to steer the discussion back on track if it appears to be heading off down a less promising avenue. Encouraging a respondent to elaborate further on something (as in the case illustrated below) mentioned in passing during a lengthy discussion is a skilful technique, which supports a naturalistic style of interviewing, as the following demonstrates:

PD: You mentioned before that you had done some benefit book stuff?
F: Yeah – wor we used to do was . . .

Again later after a lengthy description of shoplifting, more detail was sought on security obstacles – an issue raised by the interviewee but glossed over. Further prompting of that issue is achieved by repeating the words of the respondent:

PD: You mentioned security tags and foil?
F: It depended on the shops – some shops we would just . . . when we just started out we didn't know about the foil so we used ter take in a pair of wire clippers, take them off and dump them in the changing room but they made a hell of a crack – have you heard wire being cut? It made a hell of a bang so we just used ter take them off an shove them anywhere – yer know what I mean – somewhere they wouldn't be found like in a pocket of someone else's coat and . . .then we found out about the foil an we started doin that.
PD: How did you find out about that?

An experienced interviewer is able to encourage a natural narrative along restricted topics using the semi-structured interview schedule in order to introduce the topics. Equally important is the ability to trigger women's memories of particular events, as the following extract shows:

PD: You mentioned before that you used to go to the Metro Centre . . .
F: Oh yeah right there was this one time, it was Christmas time right . . .

In the majority of interviews – and all of those in prison – contemporaneous notes were taken with a small number being tape-recorded. Note taking in an interview is neither easy nor unobtrusive. Both parties' use of language as well as the nature of the accounts being offered complicates this. Interviewees' language and subject matter may be humorous or alternately violent, distressing and shocking. The interviewer's language can be adapted to suit that of the interviewee by repeating some of the words and phrases used by the women. Sometimes direct questions or comments may need to be further explained or clarified. The following extract from the same in-depth interview illustrates how respondents use their own vernacular but also shoplifting slang:

> *F*: So me an Mandy was in the aisle with the trolley and Sandra was at one end of the aisle and Kelly was at the other and we were like walkin round to see if we could see if there was any walkers in and we used to have like this little code we used ter go like if there was a walker in we used ter go . . . 'these boots are made for walkin' ..(singin). . . ..
> *PD*: Walkers – you mean like security?
> *F*: Store detectives and that's how we knew when there was one in . . .

Prison or street slang has to be quickly absorbed and used. Sometimes specific phrases and words used by the women are the only ones that properly encapsulate the meaning and description of events, and these are worth recording verbatim. Other times it is sufficient to record the gist of the account and a balance has to be struck between stemming the flow of the discussion, maintaining it, and getting it all down on paper. It is useful to adopt a method of marking which are the précised bits of the interview and which are the direct quotes, as these are often very difficult to decipher after the interview. Whilst language matters, the content must not be forgotten. Relevant details must be recorded without missing anything that could be further explored as more qualitative data. It is important to strike a balance between reacting naturally to disclosures whilst not appearing too shocked. Discussions inevitably involve some distressing stories about childhood relationships, backgrounds and circumstances; some of this may be expected but nevertheless it might be distressing both to disclose and to hear about it. On occasions such experiences of interviewing may have effects on the interviewer as researcher in the longer term. Secrets and confidences divulged by the interviewee can have a traumatic effect on researchers who are unprepared for them or who have built up a relationship and rapport with a female interviewee.

Validity revisited

Some issues concerning validity and reflective practice have already been mentioned but in presenting research findings in a final report, book or article further validity issues need to be addressed. The vast majority of those interviewed within the prison suffered from an addiction to drugs – mainly speed, amphetamines, heroin and cocaine – and to alcohol. I was both surprised and shocked to discover the high proportion of female inmates who had a drug problem despite knowing that an increased number of women received into prison have been sentenced for drug-related offences (Devlin, 1998), and that prisons generally report drug use as a major problem. Although drugs offences were not generally the reason for their most recent sentence, many of the women I interviewed were close to, or involved with, drug use. This complicated their patterns of offending, exacerbating the extent of it, and it also affected the way in which some of the interviews were conducted. The provision of cigarettes to smoke during the interview, to some small extent helped calm and compose the women, who often found concentration and conversation difficult to maintain.

It is not unknown for some commentators to denigrate research conclusions based on interviews with known criminals (see for example Jupp, 1989: Chapter 4). The research described here also relies on self-report data provided by offenders. In self-reports there is always a risk of under-reporting or of exaggeration. The fact that some of the women discussing their experiences of committing crime were drug users invites criticism of the research. Some women interviewed in prison did exhibit signs of restlessness and an inability to concentrate but the majority were keen to talk and be listened to. The discussions primarily focused upon the manner in which they carried out their activities, the locations in which these took place and details of the methods and networks they were a part of. Little of the information could be exaggerated. The greatest problem experienced was in finding out the detail of the experience. Stealing and 'grafting' had become such a way of life for many of the women that they took it for granted that the methods and techniques employed to carry out their crimes were common knowledge and a matter of routine. Although the women interviewed had in the main been labelled dishonest by the police and courts, there is no reason to suggest that their stories and narrative accounts were fabricated.

After interviewing

It is good practice to transcribe and write up interviews within one or two days after the interviews have taken place. In this way the nuances of the conversation that might not be written down or captured on tape can be remembered more accurately and also noted as reflective data. Writing up naturalistic-style interviews is a time-consuming and consequently costly activity and a well-planned research project will need to take account of this. It may also be useful to record extra details surrounding the interviews such as where and when they took place, what else was going on in the background (alarm bells, the clip clop of female officers' heels in corridors and shouting in prison), who else was around (neighbours calling, children and babies' presence at women's homes) and other contextual but factual details. Other thoughts about the interviews are also likely to be on the researcher's mind. Transcribing discussions, words and facts is part of the interviewing process but so too is the writing up of these other thoughts. They are part of what is often called reflective practice. Reflecting on the experiences of interviewing is natural and ongoing and it is useful to adapt the use of a research diary to record these 'extra' ideas and thoughts, concerns and feelings. This may appear to end the process of doing interviews, but this is an emotionally and physically draining form of interaction and upon returning home from interviews the researcher may not only be exhausted but may also experience a feeling of anticlimax. Immediately to commence the time-consuming task of writing up is not realistic. After a period of being alone subsequent to interviewing, there is a need to get the interviews down on paper or talked about. Talking about the interviews without disclosing confidences or compromising anonymity is a useful practice and serves several functions, including diffusing feelings of stress and euphoria after the intensity of the interview situation as well as preparing for their analysis and discussing fieldwork with colleagues and supervisors.

Conclusion

In describing the planning and conducting of interviews with female offenders, a number of themes arise. First, research of this nature can be understood as a continuous process of decision-making. Second, a variety of research methods and tools can be employed whilst the 'rules' of good research practice suggest which form of interviews are best suited to the particular research questions being investigated. Third, original and imaginative research often derives from different combinations of the above so that the perspective preferred by the researcher might shape the research in a particular way, for example in order to develop a feminist perspective. At various stages of the ongoing process of decision-making in relation to the research, a specific perspective might be developed and good research practice maintained through careful consideration of the following: access issues, sampling methods, interviewing skills, politics, ethics, reflexivity and validity.

An enjoyable piece of qualitative research must also be effective. Approaching the fieldwork as a continuous process of decision-making encourages the researcher to reflect continuously upon what is being done in a particular context and why, and on how this relates to the research aim and research questions. There are many different ways in which female offenders might be the subjects of a research inquiry and there is never a single best way of doing this. Methods used might involve deep surveys and/or the use of official statistics and/or interviews. There are always decisions and trade-offs to be made in fieldwork-based research, which in turn affect the research experience and outcome. In this research the most appropriate way of investigating women who commit crime for economic gain was to use qualitative semi-structured interviews.

Suggested readings

Carlen, P. (1988) *Women, Crime and Poverty.* Milton Keynes: Open University
Press.
Maher, L. (1997) *Sexed Work. Gender, Race and Resistance in a Brooklyn Drug Market.*
Oxford: Clarendon Press.
Mason, J. (1996) *Qualitative Researching.* London: Sage.
Sapsford, R. and Jupp, V. (1996) *Data Collection and Analysis.* London: Sage in
Association with the Open University.

References

Cain, M. (1994) 'Realist philosophy and standpoint epistemologies or feminist
criminology as a successor science', in L. Gelsthorpe and A. Morris (eds) *Feminist
Perspectives in Criminology.* Milton Keynes: Open University Press.
Campbell, A. (1991) *The Girls in the Gang,* 2nd edition. Oxford: Basil Blackwell.
Carlen, P. (1988) *Women, Crime and Poverty.* Milton Keynes: Open University Press.
Carlen, P., Christina, D., Hicks, J., O'Dwyer, J. and Tchaikovsky, C. (1985) *Criminal
Women.* Cambridge: Polity Press.
Cohen, S. and Taylor, L. (1972) *Psychological Survival: The Experience of Long-Term
Imprisonment.* London: Penguin.
Daly, K. (1994) *Gender, Crime, and Punishment.* New Haven and London: Yale University
Press.
Davies, P. (1999) 'Women, crime and an informal economy: female offending and crime
for gain', in *British Criminology Conferences; Selected Proceedings Volume 2*
[hhp;\\www.lboro.ac.uk\departments\ss\bsc\bccsp\vol02\\01davie.htm].
Devlin, A. (1998) *Invisible Women.* London: Waterside Press.
Field, S. (1990) *Trends in Crime and Their Interpretation.* Home Office Research Study
No. 119. London: Home Office.
Gelsthorpe, L. (1994) 'Feminist methodologies in criminology: a new approach or old
wine in new bottles?' in L. Gelsthorpe and A. Morris (eds) *Feminist Perspectives in
Criminology.* Milton Keynes: Open University Press.
Gelsthorpe, L. and Morris, A. (eds) (1994) *Feminist Perspectives in Criminology.* Milton
Keynes: Open University Press.
Glaser, B. and Strauss, A. (1967) *The Discovery of Grounded Theory.* Chicago, IL: Aldine.
Hammersley, M. and Atkinson, P. (1983) *Ethnography: Principles in Practice.* London:
Tavistock.
Heidensohn, F. (1985) *Women and Crime.* London: Macmillan.
Hudson, A. (1994) '"Elusive subjects": researching young women in trouble', in
L. Gelsthorpe and A. Morris (eds) *Feminist Perspectives in Criminology.* Milton
Keynes: Open University Press.
Jupp, V. (1989) *Methods of Criminological Research.* London: Allen and Unwin.
Maher, L. (1997) *Sexed Work. Gender, Race and Resistance in a Brooklyn Drug Market.*
Oxford: Clarendon Press.
Marsh, C. (1988) *Exploring Data: An Introduction to Data Analysis for Social Scientists.*
Cambridge: Polity Press.
Mason, J. (1996) *Qualitative Researching.* London: Sage.
Mayhew, P. (1996) 'Researching crime and victimisation', in P. Davies, P. Francis and

V. Jupp (eds) *Understanding Victimisation*. Newcastle: Northumbria Social Science Press. Chapter 3.

Maynard, M and Purvis, J. (eds) (1994) *Researching Women's Lives from a Feminist Perspective*. London: Taylor and Francis.

McCracken, G. (1988) *The Long Interview*. Beverly Hills, CA: Sage.

Naffine, N. (1997) *Feminism & Criminology*. Cambridge: Polity Press.

Oppenheim, A.N. (1992) *Questionnaire Design, Interviewing and Attitude Measurement*. London: Pinter.

Sapsford, R. and Jupp, V. (1996) *Data Collection and Analysis*. London: Sage in Association with the Open University.

Stanley, L. and Wise, S. (1993) *Breaking Out Again: Feminist Ontology and Epistemology*. London: Routledge.

Steffensmeier, D. and Allen, E. (1996) 'Gender and crime: toward a general theory of female offending', *Annual Review of Sociology*, 22: 459–487.

Walklate, S. (1996) 'Can there be a feminist victimology?', in P. Davies, P. Francis and V. Jupp (eds) *Understanding Victimisation*. Newcastle: Northumbria Social Science Press. Chapter 2.

5

DOING REALISTIC EVALUATION OF CRIMINAL JUSTICE

Nick Tilley

Contents

Evaluation studies are at the applied end of criminological research. They are principally concerned with informing policy and resource allocation decisions, by examining how programmes work out in practice. In criminal justice they represent part of an effort to ground what is done in evidence rather than prejudice or wishful thinking. Evaluation studies might, on this account, seem worthy but dull affairs: rather technical exercises of little or no theoretical interest or significance. Whilst, sadly, there certainly are plenty of dull and pedestrian evaluation studies, the most informative studies do use, develop and test theory, and necessarily so. They can certainly be technically difficult, but high quality, truly worthy realistic evaluations also offer interesting, important and inescapable theoretical challenges.

Most, if not all, interventions in criminal justice are concerned with changing human behaviour. They are intended to make a difference. It is hoped that one or more person will behave other than they would have done without the intervention. Whereas most science is concerned with identifying and understanding regularities, evaluation research is fundamentally concerned with change, and with efforts to engineer it in a predictable and consistent way.

There are various approaches to evaluating interventions (see Chelimsky and Shadish, 1997; Shadish et al., 1991). The gold standard for outcome-based evaluation has come to be the random controlled trial. In the medical field this has even been referred to as a 'shibboleth' (Susser, 1995). In the case of community-level interventions, the classic random controlled trial is impractical – think about the difficulties and costs in randomly (and blindly) allocating large numbers of communities to treatment and non-treatment. Thus, 'quasi-experiments' are run instead (Cook and Campbell, 1979). These, at their best, involve choosing experimental and control communities on the basis of measurable initial similarities; making 'before' measurements in both; applying the intervention to the experimental site (or sites); making 'after' measurements again in both the experimental and control sites; and finally comparing changes to see whether the experimental has outperformed the control.

There have been criticisms of the use made of the quasi-experimental method (Connell et al., 1995; Pawson and Tilley, 1994, 1997). 'Realistic evaluation' describes an effort to formulate an alternative evaluation methodology. This retains a commitment to scientific method and to efforts to maximize objectivity. It is, thus, not a retreat from the serious ambitions of the architects of quasi-experimental evaluation to provide a robust evidence base for developments in policy and practice (Campbell, 1969). It just construes this effort in different and, we think, better ways (Pawson and Tilley, 1997).

Realistic evaluation and its rationale

Understanding how interventions have their effects

It is sometimes easy to see how an intervention will have its effect. If a person is hanged then they can commit no more crimes. If a person is incarcerated securely, while inside they simply cannot commit external crimes directly. With regard to crime prevention, sufficient physical security may make some targets impossible to reach for all practical purposes. In these cases we can see straight away how the one set of measures prevents the criminality, and the other prevents victimization that could otherwise be expected.

Interventions to change behaviour, however, generally work in less direct ways. Improvements in lighting, closed circuit television (CCTV), property marking, prison education, probation, Neighbourhood Watch and so on do not change criminal behaviour or victimization through direct incapacitation or physical protection. Instead, they do so by some other, indirect means – by in some way affecting the decisions or responses of those whose behaviour alters. Those leaving prison, for example, are not directly constrained from crime by the education they have received. Marked property is no less physically stealable than that which is unmarked. If behaviour is made to change, it is something else about prison education or property marking that leads offenders to act differently.

Interventions also produce unintended consequences: they may make a difference that was not expected or meant. When new cars were made more difficult to steal by the introduction of steering wheel locks, those wanting to steal cars turned their attention to older ones to which steering wheel locks had not been fitted (Mayhew et al., 1988; Webb and Laycock, 1992). More happily, when (less toxic) natural gas replaced (more toxic) town gas, fewer people committed suicide because of the loss of a convenient, relatively painless method (Clarke and Mayhew, 1988).

Even where the means by which interventions have some of their effects are direct, as in the examples of hanging, incarceration and physical security, they may also have other unintended effects through less direct means. Hanging murderers, for example, may reduce the disincentive to violent resistance to arrest; incarceration may increase capacity for criminal behaviour and inhibit opportunities for non-criminal careers; and heavy physical security may be read by prospective offenders as a sign that there are potential goods to steal.

Dealing with variations in impact

Few interventions produce the same effects in all circumstances. One of the most studied criminal justice interventions in the United States has been mandatory arrest for domestic violence. Of six major studies, three found that these arrests increased the frequency of officially detected offending, and three found that they reduced it (Sherman, 1992). More generally, what is consistently found from reviews of literature on various criminal justice interventions is inconsistency of impact. This is the case even when studies with technically weak research designs are excluded (Gendreau and Ross, 1987; Martinson, 1974; Poyner 1993; Sherman et al., 1997).

Think about the circumstances affecting variations in ways in which hanging, incarceration and physical security may have side effects through indirect means. Are disincentives to resistance to violent arrest likely to operate in the same way for domestic murderers, 'mercy-killers', and gangs? Is a spell of incarceration likely to increase criminal capacity and disable future legitimate employability equally amongst old lags and those early in their offending careers? Is an improvement in physical security likely to be understood in the same way in a jeweller's, where it is already known there will be a plentiful supply of high value goods, and in a modest household?

The fact that most effects are brought about indirectly raises research questions about *how* they are produced. The fact that interventions produce a range of effects raises research questions about *diverse outcomes*. The fact that different effects are produced in different circumstances raises questions about what the *salient conditions* are that shape the variation.

'Realistic' evaluation research is precisely concerned with finding out what outcomes are produced by interventions, how they are produced, and what is significant about the varying conditions in which the interventions take place. We use the term 'mechanism' to describe the way in which an intervention produces its outcome. We use the term 'context' to describe the salient conditions for the mechanism (or mechanisms) to be triggered. We use the term 'outcome pattern' to describe the sets of effects brought about by the mechanisms triggered in salient contextual conditions (Pawson and Tilley, 1997).

Putting context, mechanism and outcome pattern together is the real trick. The product of a piece of realistic evaluation is the development or refinement of what we call a 'context, mechanism, outcome pattern configuration', or 'CMOC' for short. The CMOC is produced in answer to what we construe as the crucial question for evaluation, 'What works for whom in what circumstances?'

Problems in other approaches

Because interventions work differently in differing conditions to generate varying outcome patterns, realistic evaluation is sceptical of the value of many traditional experimental methods. These are on the lookout for invariant relationships, but we think they are doomed to fail (Pawson and Tilley, 1994, 1998). They typically disregard contextual variation, by controlling it out. They employ a caricature of real scientific method. They are apt to steamroll over variation in context, mechanism and outcome to look for a net effect, disregarding the complex (and sometimes contradictory) change patterns that can be generated. Most seriously, it is hard to find many useful lessons for criminal justice policy and practice which have followed in the wake of these evaluations.

There is a real risk that crass value-for-money studies will focus on context-blind outcome patterns from isolated quasi-experimental studies to determine which policies or practices should be implemented across the board. In order to avoid 'reinventing the wheel', evaluations are expected to identify successful practices that can be taken off the peg and implemented with confidence in their (cost-) effectiveness. In practice, this is likely at best to lead to the fitment of wheels that squeak because they do not properly fit the problems; more seriously, the effect may be to make matters worse, as in the example of mandatory arrest for domestic violence.

First steps in realistic evaluation in criminal justice issues

Realistic evaluation before 'realistic evaluation'

Several evaluations in criminal justice had implicitly adopted a broadly realistic approach before Ray Pawson and I wrote about it. Gloria Laycock discusses how property marking prevented burglary in the specific contexts (and methods of implementation) of the South Wales villages where the demonstration project was introduced (Laycock, 1992). In the small, relatively isolated villages, it was the high density implementation of property marking and extensive publicity that dissuaded the (overwhelmingly local) offenders from burglary, by convincing them that risks of apprehension had increased; it was not the property marking *per se* that did the trick.

An early 'simple' example: mechanisms at work

Armed with ideas about realistic evaluation Pawson and I had already mapped out in outline (Pawson and Tilley, 1992), I arrived in the Home Office as a research consultant in 1992. My job was to help with Safer Cities evaluations of suites of schemes with some common theme running through them, where sufficient data had already been collected or could be collected. It happened that several schemes in different cities had involved the introduction of CCTV in car parks. Moreover, before-and-after data were often available, or could be collected.[1] I began thinking about car parks, crime and CCTV in realistic terms, and tried to forge a theory. I came up with the following possible mechanisms (Tilley, 1993a):

1 CCTV reduces car crime by making it more likely that present offenders will be caught, stopped, removed, punished and deterred.
2 CCTV reduces car crime by deterring potential offenders who will not wish to risk apprehension and conviction by the evidence collected on videotape or observed by an operator on a screen on which their behaviour is shown.
3 The presence of CCTV leads to an increase in usage of car parks, because drivers feel less at risk of victimization. Increased usage enhances natural surveillance, which deters potential offenders, who feel they are at increased risk of apprehension in the course of criminal behaviour.
4 CCTV allows for the effective deployment of security staff/police in areas where suspicious behaviour is occurring. They then act as a visible presence, deterring potential offenders. They may also apprehend offenders red handed and disable their criminal behaviour.
5 The publicity given to CCTV and to its usage in catching criminals is received by potential offenders, who avoid the increased risk they believe to be associated with committing car crimes in car parks. The perceived risks of offending exceed the perceived benefits, and offending either ceases or is displaced by place or offence.
6 CCTV, and signs indicating that it is in operation, symbolizes the effort to take crime seriously and to reduce it. The potential offender perceives crime to be more difficult or risky and is deterred.
7 Those car crimes which can be completed in a very short space of time will be reduced less than those which take more time, as the offender calculates

the time taken for the police or security officers to come or the probability that panning cameras will focus on him/her.[2]

8 CCTV and notices indicating that it is in operation remind drivers that their cars are vulnerable, and they take greater care to lock them, to operate any security devices, and to remove any easily stolen items from view.

9 Cautious drivers, who are sensitive to the possibility that their cars may be vulnerable and are habitual users of various security devices, use and fill those car parks that have CCTV and thereby drive out those who are more careless, whose vulnerable cars are stolen from elsewhere.

The influence of variations in context

I also thought about variations in context that might be salient for the activation by CCTV of different mechanisms:

- A given rate of car crime may result from widely differing prevalences of offending. For example if there are 1,000 incidents per annum, these may be committed by anything from a single very busy offender to as many as 1,000 offenders, or still more if they operate in groups. A mechanism of disablement (as in [1] above) holds potential promise according to the offender–offence ratio.
- A long-stay car park may have an enormous influx of vehicles between 8 and 8.30 in the morning, when it becomes full up. It may then empty between five and six in the evening. If the dominant CCTV mechanism turns out to be increased confidence and usage (as in [3] above) then this will have little impact because the pattern of usage is already high, with little movement, and is dictated by working hours not fear of crime. If, however, the car park is little used, but with a very high per user car crime rate then the increased usage mechanisms may lead to an overall increase in numbers of crimes but a decreased rate per use.
- Cars parked in CCTV blind spots in car parks will be more vulnerable if the mechanism is increased chances of apprehension through evidence on videotape (as in [2] above), but not if it is the changed attributes/security behaviour of customers (as in [8] or [9] above).
- The local patterns of motivation of offenders, together with the availability of alternative targets of car crime, furnish aspects of the wider context for displacement to car crimes elsewhere, whatever crime reduction mechanisms may be fired by CCTV in the specific context of a given car park.
- In an isolated car park with no security staff and the police some distance away, the deployment of security staff/police as a mobile and flexible resource to deter car crime (as in [4] above) is not possible.

This describes some realistic theorizing about CCTV, car parks and car crime, though it can be acknowledged that further mechanisms and contextual variations might be important. What the example shows is that with something as apparently simple and mechanical as CCTV and car parks, contextual variation needs to be considered in order to understand what mechanisms may or may not be triggered to produce quite widely varying outcome patterns. Moreover, it also pointed to some of the sorts of data that would need to be collected to test or arbitrate between varying theories about context, mechanism and outcome pattern configurations. Without the theory the researcher would not be directed to look at car park usage patterns, changes in selection of car park, visible goods left in the car, the distribution of locations in the car park where car crime take place and so on. It might be thought that a car park is a car park is a car park. Some relatively simple theoretical work is needed to figure out possible salient ways in which they vary. Moreover, in the case of the Safer Cities schemes, notwithstanding the supply of reasonable (or cleanable) incident data, the amount of arbitration between the different possibilities laid out through the theory proved quite limited.

On reading this account of crime, CCTV and car parks, one very distinguished American criminologist wrote to me berating the simplicity of the example. Most programmes, he said, were much more complex. He was absolutely right in his remarks, though not in the implied inference. If, when examining something as simple as CCTV in car parks and its outcomes only for car crime, one is liable to be seriously misled by flattening out variations in context and mechanism, problems are liable to be more serious when one is doing the same for multi-faceted programmes implemented in more heterogeneous contexts.

The problem of replication

While at the Home Office working on Safer Cities, I was also asked to look at replications of the highly successful and widely promoted Kirkholt Burglary Prevention Project (Forrester et al., 1988, 1990). This turned out to be much more difficult than I had anticipated. I had not started out with any intention to use realism in this case, but in the end it proved invaluable in making sense of the problems.

The Kirkholt project had apparently led to dramatic reductions in burglary on the estate in Rochdale – from an incidence rate of about 25 per 100 households to one of 6 per 100 households between 1986–87 and 1989–90. Safer Cities co-ordinators had been taught about Kirkholt and were encouraged to reproduce it. Some sceptical voices had queried Kirkholt's claims to achievements. The Safer Cities work seemed to furnish an ideal basis for checking the effectiveness of the Kirkholt approach (Tilley, 1993b, 1996).

I focused on three supposed replications of the Kirkholt project within Safer Cities, where there seemed to be adequate data on what had been done and also on crime patterns. It turned out that, whilst there were some resemblances, these initiatives were different in many respects from each other and from the original Kirkholt project. The projects also varied in the measured crime pattern outcomes. Did these replications vindicate Kirkholt?

Could they all *really* be construed as true replications? Could any of them (or any parts of them) be deemed authentic replications?

I went back to the papers on Kirkholt, and to notes written when I had visited the estate and spoken to those involved in the project, to see if they would help. The following are ten attributes of the Kirkholt initiative.

1   'Kirkholt' was conceived and undertaken as a well-resourced *demonstration project*.
2   'Kirkholt' was about developing crime prevention measures in *high crime areas*.
3   'Kirkholt' was about tackling high crime areas, which are *clearly circumscribed* and can thus be treated as *identifiable communities*.
4   'Kirkholt' was about the *removal of highly attractive targets* (coin meters), which had rendered the area a popular one with burglars in which such 'money boxes' could confidently be expected.
5   'Kirkholt' was about carefully *diagnosing a particular crime problem* (burglary in an estate) and tailoring responses to this.
6   'Kirkholt' was about developing an *effective inter-agency response* to crime.
7   'Kirkholt' was about *harnessing the community* to protect itself from crime (through cocooning).
8   'Kirkholt' was about focusing on *multiple victimization* and reducing it.
9   'Kirkholt' was about *clarity of initial research, clarity of crime prevention method tailored to research findings, and clarity of leadership* in implementing measures.
10   'Kirkholt' was about burglary prevention and was *offence specific*.

None of the putative replications included all ten attributes. None was or could be identical to Kirkholt. Different decisions had been taken about what was essential. The original Kirkholt reports described what had been done, but could not on their own unequivocally inform decisions about what could or could not be deemed a replication.

Empiricist, 'mirror' accounts of Kirkholt would not do. There were unavoidable decisions for those conducting replications about what could be deemed a similar enough problem, a similar enough setting and a similar enough intervention. Moreover, no description of the Kirkholt initiative itself could be extensive enough to capture all the detail of what was done and how it was done. Implicitly or otherwise a model of some sort – a theory – has to be used to identify the salient features for any replication.

Realism provided the necessary ingredients for such a model: specification of the crucial contextual conditions for the intervention, the change-inducing mechanisms that will be triggered by the intervention, and the anticipated outcome pattern that will be generated by triggering these mechanisms. This comprises a 'context, mechanism, outcome pattern configuration'. The lack of these within accounts of Kirkholt and their subsequent projects explains the undecidability of their adequacy as replications to test the generalizability of the Kirkholt achievements. In the end, I was forced to conclude that the original question posed (Did Kirkholt replications within Safer Cities confirm the generalizability of Kirkholt?) could not be answered in the absence of realistic theory.

In the period following the Kirkholt project Ken Pease, alongside a variety of co-workers and a growing number of other researchers, picked up, modelled and generalized the repeat victimization element of Kirkholt (see Pease, 1998). What we see is indeed a refinement/cumulative understanding of repeat victimization CMOCs (see Pawson and Tilley, 1997).

The problems in decision-making about Kirkholt replications will, of course, be found in any other intervention also. The practitioner cannot sensibly decide what can appropriately be replicated without a realistic model as a guide. Realism suggests that the evaluator needs to elicit, construct, test and refine those models recursively through a series of projects.

Getting going in realistic evaluation

Let us look at how to do realistic evaluation. Take the issue of mandatory arrest for domestic violence. A good starting point is to do some thinking! In a couple of workshops Ray Pawson and I have given participants an exercise on it. We have done this with generalist evaluators and with educational social workers, not with criminologists or experts in the criminal justice system. Participants were thus unlikely to be privy to the literature, or to work with victims or perpetrators of domestic violence. Following a session on realistic evaluation methods, we presented them with the following exercise:

> You are tasked with evaluating a programme of automatic arrest following calls for service to scenes where men have been violent to their partners. The police are instructed to make an arrest, though this does not necessarily lead to prosecution. Consider the following questions and complete the attached form:
>
> How might mandatory arrest reduce repeated domestic violence (possible 'mechanisms' triggered)?
> What might be differing sub-groups amongst whom different mechanisms are triggered (variations in 'context')?
> How might outcomes vary amongst different sub-groups?
> How would you set about testing your CMOC theories?
> What data would you need? How would you collect it?'

We asked workshop participants to brainstorm and complete a simple form. Table 5.1 shows what they came up with.

These are conjectures. They are not informed either by work experience or by familiarity with the relevant literatures. Several, or even all may be quite mistaken. What is interesting is that they include, among other things, the patterns and possibilities raised in Sherman's insightful *post hoc* efforts to

Table 5.1 *Context, mechanism, outcome conjectures from realistic evaluation workshop*

Mechanism	Context	Data to test expected outcome pattern
Women's shame	Membership of 'respectable' knowing community	Reduced levels of reporting of incidents amongst those with close attachments to communities valuing traditional family life
Women's fear of recrimination	History of violence; culturally supported violence; alcoholism of offender	Reduced levels of reporting incidents amongst chronically victimized
Women's fear of loss of partner	Emotional or financial dependency on partner	Reduced level of reporting amongst poorer and emotionally weaker women
Women's fear of children being taken into care	Pattern of general domestic violence against whole family	Reduced level of reporting amongst families known to social services
Women's empowerment	Availability of refuges; support for women; financial resources of women	Increased levels of separation where support and alternative living arrangements available
Incapacitation of offender	Length of time held	Short-term reductions in repeat incidents
Offender shame	Membership of 'respectable' knowing community	Reduced repeat violence within 'respectable' communities
Offender anger	Cultural acceptability of male violence to women; what man has to lose from brushes with the law	Increased levels of violence amongst those violence-sanctioning communities marginal to mainstream society
Offender shock	Offender attachment to partner; self-image as law-abiding respectable person	Reduced levels of violence, and help-seeking behaviour amongst short-tempered 'respectable' men
Changed norms about propriety of domestic violence	Positive publicity	Reduced levels of reported and unreported violence

make sense of the mixed findings from the American research. Sherman refers to the positive impact in respectable areas of high employment generated through mechanisms triggered by shaming; and a negative effect in non-respectable areas, especially amongst the unemployed, generated through anger-triggered mechanisms. Presumably, if the interpretation is correct, in an area with the 'right' balance positive and negative impacts would cancel each other out and there would be the (highly misleading) appearance of no effect at all!

What realistic evaluation offers, even using rather rough and ready and uninformed theory, is a set of conjectures for looking at internal variation in impact of a programme on specified sub-groups. This provides a test of impact and a vehicle that can be used directly to arbitrate between, refine and elaborate an account of who the programme works for, how and in what circumstances.

Learning programme theories

Of course, no real realistic evaluation would start from the theories of a bunch of folk attending a workshop. Programme architects, programme workers and programme participants are ordinarily rich sources of CMOCs, though they would never use such a convoluted term. Realistic evaluators try to forge a 'teacher–learner' relationship with those implicated in programmes to elicit CMOC theory or the ingredients of theory. This is often best done by asking individuals to talk through examples, or cases. These can be used to generate informed theory, which the evaluator can then try to test systematically.

There are also other sources of theory, not least of which is previous research, including evaluation studies of previous analogous programmes. At a meeting in Germany in 1998 to discuss experimental methods and evaluation in criminal justice, I explained briefly the sorts of ideas workshop participants came up with in relation to mandatory arrest and spousal abuse. One of the participants had taken part in the suite of studies undertaken on the issue. He responded with incredulity. He could concede, however, that armed with some theory in advance, evaluation studies could be designed to test it. What he could not believe is that a group, using no more than common sense and some structured questions, could come up with a range of ideas, including those resulting, *post hoc*, from America's most frequently evaluated criminal justice intervention. They did. Moreover, in doing so they showed how theory-building, even by the substantively inexpert, in advance of evaluation could inform targeted measurement to test ideas about how programmes work for whom in what circumstances. What is aspired to in realistic evaluation is something more valid and useful than *post hoc* efforts to make sense of mixed findings.

Doubts about realistic evaluation

Realistic evaluation's criticisms of quasi-experimental evaluation methods have been attacked by the authors of those papers on which Pawson and I have focused in developing our arguments (Bennett, 1996; Farrington, 1998). It is up to readers to make up their own minds about these debates. It will be clear that Pawson and I have not found the counter-arguments presented credible. In this section, I respond briefly to a range of other thoughtful comments that have been made about realistic evaluation.

There is nothing new in realistic evaluation. As already indicated, evaluations following the logic of realism were certainly conducted before Pawson and I conceived the notion, even though they were not formulated as such. We have emphasized that realistic evaluation is rooted in others' philosophy of science and philosophy of social science (for example Bhaskar, 1975; Harré, 1986): what we have done is to explore how these can be applied to the practical issues of conducting evaluation studies. We have been delighted to find some parallel developments in thinking about evaluation method in the United States (notably Connell et al., 1995; Julnes et al., 1998). Some of this work has taken place independently and contemporaneously, reflecting perhaps a growing disenchantment with traditional approaches.

Experimental evaluation can also be realistic. Even amongst those who go along with much that is said in realistic evaluation about the need to attend to context and mechanism, there are serious and persistent objections to the conclusion that this entails abandonment of quasi-experimental designs (Julnes et al., 1998). In relation to street lighting and crime prevention, Painter (1995) has formulated realistic theory which she has tried to test using quasi-experimental methods. Pawson and I have explained the problems we perceive in quasi-experimentation, and how various examples fail to attend satisfactorily to the fundamental issues raised in realistic evaluation (Pawson and Tilley, 1994, 1997, 1998). There may, nevertheless, be two roles for quasi-experimentation. First, it may adduce evidence that a measure has had and therefore can have an effect, whilst disregarding the contextual needs for the effect to be achieved or the manner in which the effect has been brought about. Note that used in this way it cannot show that a measure cannot have an effect or, if it has had one, when and where it will do so again in the future. Second, quasi-experimentation may be harnessed to realistic purposes, where theory development has reached a point where context and mechanism specification is possible. Theory-driven quasi-experimentation may then conceivably be possible, in which allocation is strictly in accordance with conditions required by a developed model. Others may be able to point to conditions where theory development in criminal justice interventions has reached a point where this is possible.

Realistic evaluation side-steps crucial questions about net effects and value for money. Some questions do not make sense. A responsibility of the evaluator (or social scientist more generally) may be to indicate questions that are ill-conceived. In one piece of work, I was asked to estimate the value for money of forensic science (Tilley and Ford, 1996). Forensic science is used in very different ways in different police services, in relation to varying patterns of crime and criminality, in conjunction with varying other forms of evidence, and alongside varying approaches to crime investigation. The crime commission and investigation contexts and the potential crime-detecting mechanisms triggered by use of forensic science within them vary, and are not well understood. Moreover forensic science techniques and patterns of use are continuously evolving. There is and can be no fixed value for money return from varying levels of expenditure on forensic science, because there is no fixed net effect from given levels of usage or non-usage. We deceive if we pretend there is. A more sensible research agenda revolves around trying to understand how and in what circumstances (costed) forensic science can be (better) used to help attain public objectives.

Ministers and policy-officials are wedded to the quasi-experimental approach and will believe nothing else. Quasi-experimental studies can produce evidence with strong rhetorical value. They may persuade officials and ministers and also help ministers persuade others of the evidence base for decisions and policies. On some occasions, ministers have liked to produce simple slogans: for example, 'Prison works!' or 'CCTV works!' Moreover, policy-makers and practitioners often do ask 'what works' questions, such as: 'Does CCTV work? or 'Does Neighbourhood Watch work?' I recall one middle-ranking civil servant asking (exasperatedly), 'Well, does "Kirkholt" work? I wish these academics would stop arguing over it. The question is a simple one!' Notwithstanding this, we do not believe that preserving poor or misleading methodology is a good idea simply because it suits paymasters. There may, though, be a need for some awareness-raising. Experience does not suggest that ministers and officials are incapable of understanding why an alternative approach is needed. Moreover, even when bent on conveying political messages, they can say with (almost) the same rhetorical force that some measure *can* produce the intended impact, and should therefore be supported. Indeed, this is what ministers often do say.

Realistic evaluation will take too long. Traditional experimental evaluations can sometimes come out with a relatively quick (if often expensive) finding. This, though, is not always the case. The most impressive of experimental work, probably exemplified in the overall Safer Cities evaluation (Ekblom et al., 1996), took about eight years to complete. Individual realistic evaluations need take no longer than the relatively quick quasi-experimental, and may be much cheaper. What the realist position asserts, however, is that one-off answers to evaluation questions are generally not very valuable or interesting. In natural science, though crucial experiments are often referred to, matters are seldom clinched quickly. Theory development, development of procedures for testing, interpretation of findings, refinement of measurement, refinement of theory and so on, are all time consuming. Realistic evaluation calls for programme development and evaluation through which understanding improves and measures are learned about so that they can be implemented appropriately in contexts in which their prospects are good. In the world of policy, it has to be conceded, however, that decisions will seldom wait till the social scientist (or community of social scientists) comes to a consensus following a series of studies. Though policy development might in principle be more sharply focused following realistic evaluation, in practice it is more likely that *post hoc* adaptations can be made following concurrent programmes of realistic evaluation.

Realistic evaluation opens the door to the too quick and too dirty. This is the opposite point to the one made immediately above. It might appear that the critic cannot have it both ways! They can, though. The point is that individual realistic evaluations may be quicker (though not dirtier) than their experimental counterparts, whilst the derivation, testing and refinement of CMOCs is likely to take rather longer.

Realistic evaluation drives the researcher towards useless detail. There is little to be learned from findings that are bound to individual cases. Practitioners can often talk convincingly about what they did in relation to an individual which brought about a change, and why and how the intervention produced its effect. They are apt to talk about family circumstances, and critical events in the person's life such as health, employment, friendships gained and lost and so on. Though the language is not used, in effect we find that practitioners are describing series of unique CMO case studies. Try chatting to practitioners about their work, and you will soon discover that the accounts can be quite easily reconstructed in CMO terms. Those descriptive particulars may have enormous personal interest. They may also, of course, be wrong – strictly they are conjectures for which the practitioner may have more or less direct evidence. Whatever their validity, in and of themselves they are of little value to the policy-maker or to other practitioners. What the policy-maker needs is something of a more general kind, something which is predictive rather than simply a *post hoc* rationalization. It is for this reason that Pawson and I have emphasized middle range CMOCs, echoing Merton's emphasis on 'middle range' theory (Merton, 1968) or Marshall's 'stepping stones in the middle distance' (Marshall, 1963). These are more abstract and generalizable than the specifics of individual cases, but more concrete and operational than pure theory. Richard Titmuss' *The Gift Relationship*, examining the mechanisms and contexts for the provision of high quality blood for transfusion, is an example of middle range theory and research, which draws on abstract exchange theory but is more general than case by case individual examples of decisions to give blood (Titmuss, 1970). In the context of a private health market blood is paid for, and those in need of cash will decide to sell it, regardless of whether they are in a fit condition to give it or the blood is in a fit state to use. In the context of socialized medicine, however, blood is given freely out of a sense of collective interest and responsibility: there is no incentive to give bad blood or to give blood inappropriately. Thus, through the decision-making mechanisms triggered in it, the context furnished by the British National Health Service, at the time Titmuss was writing, produced more and better blood as its outcome, than that produced in the private-market-dominated USA, where other decision-making mechanisms were triggered. Titmuss provides middle range health-care-related CMOCs. Merton's own *Social Theory and Social Structure* (1968) is another example of a middle range CMOC. Of course, the broader contexts in which middle range CMOCs themselves apply are not necessarily fixed. Indeed, processes of social change outwith the scope of middle range CMOCs make them perpetually precarious. In crime prevention, important work by Paul Ekblom (1997) has highlighted endogenous sources of instability in crime prevention CMOCs: offenders are wont to adapt their techniques in the face of crime prevention methods, and preventers similarly adapt to offender innovations; both tend to take advantage of new possibilities raised by technological developments.

Conclusion

I suspect there is a personal subtext to much academic writing.

In my particular case, realistic evaluation has provided a way of dealing with two sources of contemporary unease: about that aspect of postmodernism which casts doubt on the possibility of objective knowledge; and about that aspect of modernism that promises universal unconditional truths. Realistic evaluation promises something like objective understanding of contingent regularities and changes in regularities in behaviour. It seems to me to steer a course between the Scylla of relativism and the Charybdis of absolutism.

The ideas of realistic evaluation also help sustain my hopes that social science can be useful to real policy and practice decisions. I hope student readers of this volume may feel the same, and join efforts to inform future criminal justice-related policy through realistic research. Current central government emphasis in Britain on evidence-driven policy and practice (see Goldblatt and Lewis, 1998), and the steer towards the same at a local level provided by the Crime and Disorder Act 1998, will provide hitherto unrivalled opportunities. It may be fun critically to deconstruct these efforts. It will certainly not be difficult to do so. The bigger, and to my mind more worthwhile, challenge is to make a positive contribution to them.

Suggested readings

Chelimsky, E. and Shadish, W. (eds) (1997) *Evaluation for the 21st Century*. Thousand Oaks, CA: Sage.

Ekblom, P. and Pease, K. (1995) 'Evaluating crime prevention', in M. Tonry and D. Farrington (eds) *Building a Safer Society: Strategic Approaches to Crime Prevention* (Crime and Justice, A Review of Research, Vol. 19). Chicago: University of Chicago Press.

Foster, J. and Hope, T. (1993) *Housing, Community and Crime: the Impact of the Priority Estates Project*. London: HMSO.

Painter, K. and Tilley, N. (eds) (1999) *Surveillance of Public Space: CCTV, Street Lighting and Crime Prevention* (Crime Prevention Studies. Vol. 10). Monsey, NY: Criminal Justice Press (Chapters by Armitage, Smyth and Pease; Beck and Willis; Ditton and Short; Gill and Turbin; Painter and Farrington; Pease; and Phillips).

Pawson, R. and Tilley, N. (1997) *Realistic Evaluation*. London: Sage.

Notes

1. This was trickier than might seem likely at first sight. Some incidents were reported to the police but not the car parking authorities, some to the car parking authorities but not the police. Car parks are often referred to in police records by a variety of names (for example the street where they are located, the shopping centre to which they are attached, the local informal name for the car park), making collection of data there difficult. In some cases, it was difficult to determine from the record whether the incident took place in the car park itself or on the street where the car park was located. Moreover, Ken Pease has more recently collected evidence using non-obtrusive measures (piles of toughened glass in car parks) to show that levels of non-reporting of incidents can be quite high.

2. The smart reader will have realized that this is not just a mechanism, but refers already to crime context variation.

References

Bennett, T. (1996) 'What's new in evaluation research? A note on the Pawson and Tilley article', *British Journal of Criminology*, 36 (4): 567–573.

Bhaskar, R. (1975) *A Realist Theory of Science*. Brighton: Harvester.

Campbell, D. (1969) 'Reforms as experiments', *American Psychologist*, 24: 409–429.

Chelimsky, E. and Shadish, W. (eds) (1997) *Evaluation for the 21st Century*. Thousand Oaks, CA: Sage.

Clarke, R. and Mayhew, P. (1988) 'The British Gas suicide story and its criminological implications', *Crime and Justice*, 10.

Connell, J., Kubish, A., Schorr, L. and Weiss, C. (1995) *New Approaches to Evaluating Community Initiatives*. New York: The Aspen Institute.

Cook, T. and Campbell, D. (1979) *Quasi-Experimentation*. Chicago: Rand McNally.

Ekblom, P. (1997) 'Gearing up against crime: a dynamic framework to help designers keep up with the adaptive criminal in a changing world', *International Journal of Risk, Security and Crime Prevention*, 2 (4): 249–265.

Ekblom, P., Law, H. and Sutton, M. (1996) *Safer Cities and Domestic Burglary*, Home Office Research Study No. 164. London: Home Office.

Farrington, D. (1998) 'Evaluating "Communities that care": realistic scientific considerations', *Evaluation*, 4 (2): 204–210.

Forrester, D., Frenz, S., O'Connell, M. and Pease, K. (1988) *The Kirkholt Burglary Prevention Project, Rochdale*. Crime Prevention Unit Paper No. 13. London: Home Office.

Forrester, D., Chatterton, M. and Pease, K. (1990) *The Kirkholt Burglary Prevention Project, Phase II*. Crime Prevention Unit Paper No. 23. London: Home Office.

Gendreau, P. and Ross, R. (1987) 'The revivification of rehabilitation', *Justice Quarterly*, 4: 349–408.

Goldblatt, P. and Lewis, C. (eds) (1998) *Reducing Offending: An Assessment of Research Evidence on Ways of Dealing with Offending Behaviour*. Home Office Research Study No. 187. London: Home Office.

Harré, R. (1986) *Varieties of Realism*. Oxford: Blackwell.

Julnes, G., Mark, M. and Henry, G. (1998) 'Promoting realism in evaluation: realistic evaluation in the broader context', *Evaluation*, 4 (4): 483–504.

Laycock, G. (1992) 'Operation identification, or the power of publicity', in R. Clarke (ed.) *Situational Crime Prevention: Successful Case Studies*. New York: Harrow and Heston.

Marshall, T.H. (1963) *Sociology at the Crossroads*. London: Heinemann.

Martinson, R. (1974) 'What works? Questions and answers about prison reform', *Public Interest*, 35: 22–54.

Mayhew, P., Clarke, R. and Hough, M. (1988) 'Steering column locks and car theft', in R. Clarke and P. Mayhew (eds) *Designing Out Crime*. London: HMSO.

Merton, R. (1968) *Social Theory and Social Structure*. New York: Free Press.

Painter, K. (1995) 'An evaluation of the impact of street lighting on crime, fear of crime and quality of life'. Unpublished PhD thesis, University of Cambridge.

Pawson, R. and Tilley, N. (1992) 'Re-evaluation: rethinking research on corrections and crime', in S. Duguid (ed.) *Yearbook of Correctional Education*. Burnaby, BC: Simon Fraser University.

Pawson, R. and Tilley, N. (1994) 'What works in evaluation research?', *British Journal of Criminology*, 34: 291–306.

Pawson, R. and Tilley, N. (1997) *Realistic Evaluation*. London: Sage.

Pawson, R. and Tilley, N. (1998) 'Caring communities, paradigm polemics, design debates', *Evaluation*, 4 (1): 73–90.

Pease. K. (1998) *Repeat Victimisation: Taking Stock*. Crime Detection and Prevention Series Paper No. 90. London: Home Office.

Poyner, B. (1993) 'What works in crime prevention: an overview of evaluations', in R. Clarke (ed.) *Crime Prevention Studies*, Vol. 1. Monsey, NY: Criminal Justice Press.

Shadish, W., Cook, T. and Leviton, L. (1991) *Foundations of Program Evaluation*. Beverly Hills, CA: Sage.

Sherman, L. (1992) *Policing Domestic Violence: Experiments and Dilemmas*. New York: Free Press.

Sherman, L., Gottfredson, D., MacKenzie, D., Eck, J., Reuter, P. and Bushway, S. (1997) *Preventing Crime: What Works, What Doesn't, What's Promising: a Report to the United States Congress*. Available at Internet Address: http://www.ncjrs.org/works/index.htm.

Susser, M. (1995) 'Editorial: the tribulations of trials – intervention in communities', *American Journal of Public Health*, 85: 156–158.

Tilley, N. (1993a) *Understanding Car Parks, Crime and CCTV: Evaluation Lessons from Safer Cities*. Crime Prevention Unit Paper No. 42. London: Home Office.

Tilley, N. (1993b) *After Kirkholt: Theory, Methods and Results of Replication Evaluations*. Crime Prevention Unit Paper No. 46. London: Home Office.

Tilley, N. (1996) 'Demonstration, exemplification, duplication and replication in evaluation research', *Evaluation: The International Journal of Theory, Research and Practice*, 2 (1): 35 50.

Tilley, N. and Ford, A. (1996) *Forensic Science and Crime Investigation*. Crime Detection and Prevention Series Paper No. 73. London: Home Office.

Titmuss, R. (1970) *The Gift Relationship: From Human Blood to Social Policy*. London: George Allen and Unwin.

Webb, B. and Laycock, G. (1992) *Tackling Car Crime*. Crime Prevention Unit Paper No. 32. London: Home Office.

6

EVALUATING INITIATIVES IN THE COMMUNITY

Iain Crow

Contents

This chapter is based as much on personal experience of having been involved with community-based organizations for a number of years as on what textbooks say about research methodology. This is a personal experience which has been mainly as a researcher, but also as someone who has been involved with community organizations as a volunteer worker, management committee member and chair, and as a manager of a community-based initiative. So I have seen evaluation from different perspectives. On the basis of this experience there are two points that I would immediately wish to make. First, we do not evaluate enough. We do not evaluate many initiatives from which we could learn a lot, so we lose much valuable knowledge. There may be a recognition that new initiatives should be evaluated, but the money to do it properly is often begrudged, as though it is a bit of a luxury rather than a prerequisite for further development. Instead it often becomes tacked on as 'monitoring', frequently carried out by hard-pressed project staff, who have limited experience of such matters, and limited time available. This happened, for example, with many of the projects funded by the Safer Cities programme in various parts of the country. Invariably the 'monitoring' is either unsatis-factory, or at best a form of social accountancy – merely counting numbers of clients, and so on.

The second point is that there are also many initiatives which should *not* be evaluated. A lot of projects are not doing anything that is particularly novel, or have not got their act together enough to make evaluation worthwhile. Community-based projects have to put a lot of time and energy into short term fund-raising simply to survive, and they have learned that one way of increasing their chances of doing this is to try to hitch their wagon to some evaluative research: being evaluated in itself gives a project a certain amount of credibility. Other initiatives may be worth evaluating, but not yet. I learned this lesson some years ago when studying a couple of projects for rehabili-tating offenders, called multi-facility schemes (Crow et al., 1980). For two years we watched the schemes struggling to establish themselves, alas without success. While it is eminently sensible and worthwhile to undertake research which studies the formative stages of a community-based programme, it is important to be clear that this is not necessarily the same as evaluation, and a judgement has to be made about whether and when evaluation is appropriate.

In addition, the experience of evaluation I have gained, as researcher and participant, has made me aware of the gap that exists between the traditional scientific model of evaluation and the viewpoint of those who are involved in community-based initiatives. For one thing the language is different. Researchers talk about monitoring, evaluation, effectiveness, process and outcome variables: terms which are cool and calculating. Those who work in community-based agencies and organizations have to consider performance indicators, national standards, value for money: terms which evoke manage-rialism and control. It is my contention that differences between the scientific evaluator and the practitioner in terms of orientation and approach stem in large measure from the model of evaluation which is commonly adopted by social researchers.

The theory of evaluation

This section starts by referring to the traditional model of evaluation. This is based on a natural science analogy, which sees society as a kind of laboratory. However, behind laboratories are theories, from which are derived hypotheses regarding what one may expect to find as a result of empirical inquiry. Data are collected and the observed data compared with what is expected on the basis of the hypotheses. Depending on the results, theory is rejected or modified. So in the classic scientific paradigm of evaluation an experiment is set up to collect the data necessary to test hypotheses. In the medical sphere, this usually involves some form of treatment, and a new drug will be subjected to clinical trials in which some patients are given the drug and others are not, on a random basis. This is the model (or at least a crude characterization of it) usually adopted in the social sciences for evaluating interventions, although it is seldom possible to match laboratory conditions.

Various methodological and statistical techniques are employed in an attempt to ensure that like is compared with like, and that all variables except the experimental, or dependent, variable are controlled in some way.

The practice of evaluation

How does this model or theory compare with the reality of evaluation? I would suggest that the first and major problem with the practice of evaluation in criminology is that the research may be undertaken in too limited a manner. There is too little regard for the research process as a whole and theory in particular. Various approaches to evaluation may be adopted (see Everitt and Hardiker, 1996), but the model characteristically used to represent the traditional evaluation approach is portrayed in Figure 6.1.

	Pre-test	Treatment applied	Post-test
Experimental group	O_1	X	O_2
Control group	O_1		O_2

Figure 6.1   *The traditional evaluation (OXO) model*

This can be found in various forms in research methods texts, and it is used by Ray Pawson and Nick Tilley (1994, 1997) to present a critique of the 'quasi-experimental paradigm'.

This model focuses on a very narrow part of the research process, the part that concerns itself with the particular research techniques being used, as though what go into and what come out of the evaluation are 'givens', already determined. In the natural science paradigm described earlier an experiment or treatment is initiated on the basis of a theory about what is likely to work. To use the medical analogy, new drugs are not concocted at random, but on the basis of investigation of a disease and a hypothesis about what is likely to work. Although the limitations of a medical analogy must be acknowledged, it is relevant in the present context. When it comes to social intervention this *should* translate into a theory about the nature of the problem being addressed. Intervention and its accompanying evaluation are therefore the empirical test of a theoretical proposition. The model may be illustrated as shown in Figure 6.2.

THEORY ⟶ INTERVENTION ⟶ OUTCOME ⟶ MODIFIED THEORY

Figure 6.2   *Process for the revision of theory*

Alternatively, one could present this in the form of the widely accepted model, found in several texts (e.g. Rose, 1982: Chaper 2; Bryman and Cramer, 1990: 3), shown in Figure 6.3.

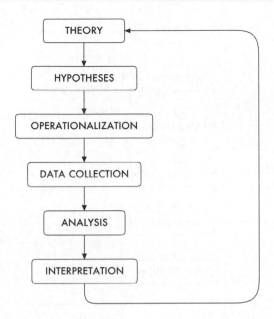

Figure 6.3   *The research process 1*

The individual components of the model in Figure 6.3 (not least the term 'theory') could be considered at length, and therefore any attempt to define them here is bound to be limited. But some explanation is necessary. Theory here means an attempt to make sense of interconnected relationships so that the empirical world can be understood in a systematic manner, and hypotheses are specific working statements about particular relationships contained within such theories. For example, a hypothesis may be stated that the criminal justice system deals with unemployed people differently from the way in which it deals with those who have jobs. This hypothesis can be tested by an empirical study which looks at whether or not employment status makes a difference to the way cases are dealt with by the courts.[1] The hypothesis may then be rejected, amended, or accepted. The framing of such a hypothesis can be seen to have its origins in theories about the ways in which, over several centuries, the state has sought to control and respond to worklessness. Testing the hypothesis depends on being able to come up with certain measures which render it susceptible to empirical inquiry. This is what is meant by the rather inelegant term 'operationalization'. For example, to test the above hypothesis it is necessary to have clear definitions of 'employed' and 'unemployed'. It may also be necessary to come up with a scale of 'sentencing severity' and to be able to take account of other, intervening variables which can affect the outcome of a case apart from employment status, such as the seriousness of the offence and the nature and extent of previous offending.

It must be emphasized that the above model is very much an 'ideal type'. Researchers may also work from 'grounded' data (Glaser and Strauss, 1967; Punch, 1998: Chapter 8) towards the development of concepts and theoretical

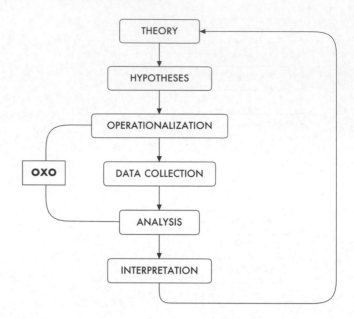

Figure 6.4   *The research process 2*

perspectives, and in practice many researchers also jump around quite a bit. This model is not peculiar to quantitative researchers; it would be recognized across a wide spectrum of research approaches, including documentary work (Cooper, 1989) and ethnography (Agar, 1986). The OXO diagram – see Figure 6.1 – can then be located *within* this broader model, as represented in Figure 6.4.

In other words the OXO model so often concentrated on by evaluators is only part of a wider process; to ignore this is to attenuate the research design.

Many projects are set up to deal with crime and offenders, and evaluation of them must take account of the process of research as a whole, rather than as a narrowly conceived technical task. To give an example, if a project is established whose expressed aim is to rehabilitate drug addicts, then evaluating it involves rather more than comparing those who go to the project with a group of addicts who do not. Inherent in the project is a theory about the nature of drug addiction and what needs to be done about it. Thus, a residential project for drug addicts such as Phoenix House has a substantial theoretical pedigree going back over many years based on the concept of the therapeutic community. Other projects may not have such a well-developed and referenced theoretical background. Some projects are set up simply as a consequence of a perceived need to 'do something' about a problem, perhaps taking advantage of a funding opportunity that has arisen. However, they will still have *some* theoretical impetus, which may be implicit within the workings of the project, rather than being explicitly stated. One can think of projects which have been set up to deal with offenders or prevent young people from becoming involved in crime – motor projects, sports projects, projects in which young people meet 'old lags' who have been to prison in an attempt to 'scare them straight', employment and training opportunities – all of which have some underlying notion of what needs to be done in response to offending, and therefore embody some kind of theory (or perhaps a cocktail of theories) about what causes people to commit crime. Much the same can be said about crime prevention initiatives, whether they are based on situational measures or attempts to prevent crime through social and economic activities, such as community regeneration and involvement. In any of these situations, an evaluation which ignores the theoretical context of an initiative and focuses only on certain measures of outcome is an incomplete piece of research which will fail to understand the process by which certain outcomes have come about. It may, however, be useful to make a distinction between the kinds of formal theories employed by social researchers and what may be termed 'practitioner theories', the notions on which those who implement programmes operate on a day to day basis.

In reality the relationship between a particular phenomenon, such as drug addiction or crime, and a specific form of intervention may be quite complex. Psychological, social and economic factors may be involved in varying degrees in an initiative. Nonetheless, whenever one encounters a programme it is important to ask what the theoretical implications are of that form of treatment or intervention. Often this is either not done, or is not apparent in a research report. This is rather like developing a questionnaire, undertaking participant observation, or using some other research technique without regard to the theoretical basis of the research. Much follows from this that affects the adequacy of an evaluation. For example, the first concern of a competent evaluator should be to give careful consideration to the theoretical basis for the initiative and its adequacy. If the theory itself is badly flawed then there is a good chance that the programme will not work. This does not mean that the theory has to be perfect: few are, and the purpose of implementing a programme is to put a theory to the test in order to reject or modify it. But, to use the words of one of my colleagues, 'Illogical theories are bound to lead to programme failure'.[2] The implications of theoretically informed evaluation cannot be pursued in detail here, but mention can be made of some of the issues relevant to the model of inquiry that is adopted.

The problems of evaluation

The first issue concerns the need to look at what an organization or initiative is meant to be doing; what its *aims* are. These are liable to vary in their degree of clarity. This may be due to a failure on the part of those responsible to articulate the aims, or it could be that they are fudged and deliberately vague because they have to meet the needs and objectives of different groups. Indeed, it is not uncommon to find that an initiative set up to do one thing (e.g. engage in youth work, or develop social skills), claims to be reducing crime amongst young people because that is how it can best get financial resources. Looked at more closely one finds the project is not doing things directed at crime reduction at all, or only tangentially. Further, some organizations have bland aims, or aims which amount to little more than trying to change the world. The judicious researcher will inquire further about what lies behind such aspirations. A clear appreciation of an organization's goals is essential because these are the key to unpacking the organization's theoretical baggage – to understanding what assumptions it is using in relation to its work.

One also has to be aware that a project may attempt to address *several aims*, and perhaps accord different priorities to those aims. Organizations may have primary, secondary, and perhaps even tertiary, goals which require investigation.

The consideration of aims also affects the *criteria* of success that are adopted. Quite simply, there is no point in measuring reconviction rates if what an organization is doing cannot be shown to be related to reducing offending. This may sound obvious and trite, but inquiries about the impact an initiative is having on offending may be irrelevant because what is happening is not in fact about crime reduction. For example, a victim–offender mediation scheme may be a very worthwhile project for many reasons, but there is a real question about whether victim–offender mediation results in crime reduction. Another example would be employment training schemes for offenders. There may be a relationship between crime and unemployment, but when evaluating employment and training programmes, such as the Community Programme and the Youth Training Scheme which operated during the 1980s, the most relevant question for those running the schemes was whether people got jobs at the end of them, not whether the schemes served the needs of criminal justice (see Crow et al., 1989: 77–83). Furthermore there can be a big gap between what the senior administrators of large programmes intend and what the staff on the ground see themselves as doing, which may also cause the researcher to reflect upon the most appropriate criteria for evaluation.

A final example relevant to the consideration of project aims comes from a study of projects providing sporting and other physical activities for young offenders. As in the case of the drug project, the immediate aim of the people running one such scheme was quite simply to motivate the young offenders to do *something* positive. This was the first and necessary step in addressing offending behaviour. But this leads one to consider the all-important means by which sports and other physical activities might have an impact on young offenders. Is it that sport occupies their free time, that it brings them into contact with new people, that it increases their social and physical skills and ultimately perhaps their employability? Or is it the role model presented by the activity leader that has an effect (see Nichols and Taylor, 1996: iii)? Similar questions may be asked of employment itself. If a job reduces the likelihood of offending is this because of the money it provides, the occupying of 'idle hands', or is it part of some wider sense of social organization? An evaluation is incomplete if it does not consider the *means* by which an outcome is achieved as well as the outcome itself, because without such an understanding it is not possible to replicate the intervention elsewhere. In other words it is not possible to achieve *programme generalization*.

This brings us to the importance of *levels* of evaluation and the dangers of a 'one model fits all' approach which seeks to find the magic bullet solution to a problem. Initiatives may take place in a variety of situations and at a variety of levels of measurement. There are one-off, single-purpose projects such as the sports programme already referred to; there are organizations whose goal is to implement change across a wide front of related activities (e.g. NACRO); there are programmes at neighbourhood, city and regional levels (e.g. Safer Cities); and there are national and trans-national levels of evaluation of the implementation of policy. Evaluation of the implementation of legislation, such as investigating the impact of the Criminal Justice Act 1991, is an example of national-level change, while one may consider that during the 1990s there was a trans-national experiment, common to the USA and Britain, in increasing the use of imprisonment to influence crime rates. Each has to be evaluated at an appropriate level, and what works at one level may not work at another. It may be inadvisable to look at whether a programme designed to increase safety in a community has reduced crime levels in an area if in fact most of the work of the initiative has reached only particular groups of individuals, and has not extended into the community as a whole. Likewise, studies of employment and training schemes for offenders have, by and large, been unable to demonstrate an impact on offending. But this could be because they failed to address, and were never capable of addressing, the wider problem of mass unemployment. The so-called OXO model, as portrayed in Figure 6.1, is one largely suited to individual treatment regimes. Although it may bear some extrapolation to other contexts, it would be a mistake to assume that it is always the right model.

There are several other features of evaluation which deserve attention, to which I can only briefly allude in this chapter. There is the all-important question of who *owns* and controls the evaluation. This is important because of what is known as programme integrity. Programme integrity refers to the extent to which an initiative retains its original aims and methods of working. If an intervention wanders away from what it was originally doing and changes direction then it is said to lose its integrity for evaluative purposes. To revert to the medical analogy, it is a bit like changing the ingredients of a new drug part of the way through a controlled trial experiment. In the case of social intervention it is not quite as simple as this. In a social programme, such as one dealing with crime and offending, some things are bound to change, perhaps because external circumstances change. The traditional model of evaluation regards this as a 'bad thing'. A more realistic approach would recognize that such changes occur and attempt to document and analyse the nature and implications of any change, while also having regard to the implications of change for outcome measures. It may be that change occurs as lessons are learned by those who run the programmes, and it may be argued that social programmes are by their very nature dynamic and evolving entities. But what does matter is the extent to which changes are recognized and managed, and by whom. An extension of this is the role played by *key individuals* and in particular whether, as happens with imaginative new

initiatives, there is a single charismatic figure, who may be the key deter-
minant of its success.

Finally, insufficient attention is often paid to what is happening to the
controls in the traditional model, whether these are individuals or areas. In
Figure 6.1 the control is represented by a blank. However, controls are seldom
if ever blank – something where nothing happens. There is usually a lot
happening in the lives of the individuals concerned, or in the control areas in
the case of a geographical initiative, and to treat them as some kind of neutral
constant in the equation is unsatisfactory. It is also a form of objectification
which, if space permitted, would give rise to further consideration of the
ethics of such research.

Although a number of separate issues have been noted, the main point is
that how they are addressed depends to a large extent on how one approaches
an evaluation, whether as an exercise which focuses primarily on controlling
extraneous variables and concentrating on outcome measures, or as a theoretically informed attempt to understand the processes at work – processes
which are influenced by the theoretical ideas which the practitioners involved
in a programme bring to it. The next section exemplifies these points by
reference to a particular instance of evaluation, the evaluation of the Communities That Care initiative, and looks at a debate that has recently focused
on that initiative.

Applying a theoretically aware approach to evaluation: the Communities That Care (CTC) initiative

Having introduced a personal perspective on the evaluation of community-
based initiatives, it is hard to ignore the fact that there has been a debate in the
UK in the 1990s about the nature of evaluation – the so-called 'paradigm
wars'. The Communities That Care (CTC) initiative has become a focus for this
debate, and is being evaluated by myself and others.

In 1994 an article by Ray Pawson and Nick Tilley was published in the
British Journal of Criminology. They argued that the quasi-experimental paradigm, sometimes also referred to as the OXO model, had resulted in moribund
evaluation, and had been 'itself a contributing factor to the 'nothing works'
lament' (1994: 291). The authors went on to submit to critical scrutiny an
evaluation by Trevor Bennett of a police initiative for reducing fear of crime,
recognizing this to be a well-executed study in OXO terms, but saying that it
failed to explain how police activities might bring about change in a community (1994: 297). It was suggested that what is needed is to understand a
programme's mechanisms and the context in which it takes place. (For a more
detailed exposition of the position taken by Pawson and Tilley see Chapter 5
in this volume.)

The original article has been the subject of subsequent exchanges in the
British Journal of Criminology between Pawson and Tilley on the one hand, and

Bennett on the other (Bennett, 1996; Pawson and Tilley, 1996), with Bennett disputing that researchers who use quasi-experimental designs 'overlook . . . mechanisms and contexts' (Bennett, 1996: 568), and denying that competent evaluators merely seek associations between treatments and outcomes. Another chapter of the debate opened when David Farrington, who was closely involved with the development of the Communities That Care programme in the UK, published an article in the journal *Evaluation* suggesting how that programme might be evaluated along fairly traditional quasi-experimental lines (Farrington, 1997), and there have been further exchanges since (Pawson and Tilley, 1998a; Farrington, 1998; Pawson and Tilley, 1998b).

I think a lot could be done to improve the way that programmes are evaluated. Indeed the approach that I have adopted in evaluation over a number of years is, I think, quite close to that advocated by Pawson and Tilley. However, what is dissimilar is that their approach is based on a very different philosophical approach to evaluation, that of 'scientific realism'. This is set out in their book (Pawson and Tilley, 1997: Chapter 3) and referred to in Chapter 5 in this volume. They explain that scientific realism involves replacing the kind of successionist thinking associated with an experimental or quasi-experimental approach to evaluation by a generative conception of causation which explores the transformative potential of phenomena, and this means having a knowledge of the context and mechanisms of any innovation (the CMOC model). While I am in favour of reform, what I find less convincing is the need for revolution. Evaluations are sometimes deficient, but this is not the same as saying that the whole paradigm is wrong. The paradigm may be inadequately implemented by being atheoretical, and by paying insufficient attention to the social dynamics of programmes and organizations. There is a principle that one does not resort to a new paradigm without having fully explored the possibilities of the one already available. Thomas Kuhn, in *The Structure of Scientific Revolutions*, talks about paradigm shift being accompanied by the awareness of anomaly between 'what should be' and 'what is' (Kuhn, 1970: 62), but also goes on to say that

> By ensuring that the paradigm will not be too easily surrendered, resistance guarantees that scientists will not be lightly distracted and that the anomalies that lead to paradigm change will penetrate existing knowledge to the core. (Kuhn, 1970: 65)

The issue is, therefore, whether we have what Kuhn describes as 'a pronounced failure in the normal problem-solving activity' (1970: 74–75). In other words, what is being suggested is that there is another option, between Pawson and Tilley on the one hand, and Bennett and Farrington on the other, which recognizes the valid criticisms that Pawson and Tilley make of certain types of evaluation, but reasserts the importance of the total research process, thus avoiding the need to invoke an entirely different doctrine.

To say all of this is not simply to be critical of the failures of other evaluators. The opportunity to undertake the kind of evaluation one would wish may not always be available because of the way in which a programme is conceived, or implemented, or funded, or a mixture of these and other factors.

One of the main reasons for becoming involved with the Communities That Care initiative is that it appears to offer an opportunity to do an evaluation which is based on previous research experience and also has a chance of being well implemented.

The Communities That Care initiative is a long-term programme attempting to build safer neighbourhoods where children and young people are valued, respected and encouraged to achieve their potential. It involves establishing a working partnership between local people, agencies and organizations to support and strengthen families, promote school commitment and success, encourage responsible sexual behaviour and achieve a safer, more cohesive community. It starts by assessing the levels of various risk factors in a community, such as family conflict, low school achievement and community disorganization and neglect, and measures outcomes in relation to drug abuse, youth crime, school-age pregnancy and school failure. Having been established in the United States (but notably not evaluated there) CTC programmes are just starting in neighbourhoods in three British cities – a kind of McDonaldization franchising of social programmes.

CTC has a number of distinctive features. It is grounded in empirical associations between risk factors and problem behaviour amongst young people, and is based on a model for evaluating change developed in the United States. It has well-defined aims and there is a clear indication of how the programme should be implemented, with those involved receiving training sessions in both the theoretical and practical requirements. The initiative proceeds by undertaking an initial risk audit, which then forms the basis for action plans. Finally, a 'detached' outcome evaluation is accompanied by a well-articulated evaluation of the processes by which outcomes are achieved. This last point will be considered further shortly.

At the time of writing, the evaluation is in its early stages, and given the advantageous features mentioned above it will be interesting to see what happens in practice. In particular I think it poses the question of whether you can treat a social programme as a homogeneous entity, like a dose of medicine administered with the same ingredients – or a beefburger recipe perhaps – or whether there will be subtle and not so subtle variations: is there one CTC programme or several? Because of the favourable circumstances that it offers, CTC does afford an opportunity to examine first and foremost whether it works, but also why it does or does not work, and it also offers an opportunity to develop an evaluation appropriate to the intervention.

What to do: implications for evaluating the CTC initiative

What are the implications of the arguments of this chapter for the evaluation of CTC? First, the evaluation needs to be theoretically aware. In the case of CTC, although there is supposed to be a theoretical framework, quite a bit still needs to be done to fully explore and articulate the theoretical context. Most importantly, there are likely to be differences between the theories held by the

American architects of the programme and the frames of reference of people on the ground in the UK cities involved.

Further, the specific components of intervention need to be examined to look at how they relate to the apparent theoretical basis for intervention, as does the process of implementation itself. Evaluation has become something of a methodological specialism, but we should not forget that it is part of a broader process of social inquiry which may need to include a combination of research methods. So while outcome variables and controls have to be considered, more than this is needed. In addition to quantifiable outcome measures, some of the research will be more qualitative than is generally envisaged in the traditional OXO model, involving individual and group interviews, observations, documentary analysis and other methods in order to record the process of change.

The deficiencies of the static OXO model are sometimes remedied by undertaking 'process research', which seeks to augment outcome research by examining the activities involved. But what exactly does process research involve? If we are not careful it may end up being no more than a descriptive account, rather than an analytic investigation. Process research must be theory-oriented. That is, it is attempting to explore the explicit and implicit theories which sustain a programme. So in addition to examining the programme theory, it should also examine how participants perceive and translate this in the course of implementation, and it is especially important to explore the personal and organizational theories which they bring to a programme – whether explicit or implicit. An example may help to clarify what is meant. With a colleague I am currently involved in evaluating the impact of 'fast tracking' procedures for persistent young offenders. This evaluative research was preceded by a study undertaken at the same youth court which sought to compare views of how to deal with young offenders amongst magistrates and clerks on the one hand, and social workers on the other (Crow, 1996). This research suggested that the two groups arrived at the issue with two contrasting practitioner theories about youth crime and how to deal with it. Put very simply, the perspective of the magistrates and clerks was based on a rudimentary form of social learning theory which led them to believe that the sooner one intervened to deal with a young offender the better, in order to deter them from reoffending. The approach of the social workers, on the other hand, tended to have been informed over the years by labelling theory. This underlines the dangers of drawing a young person into the web of the criminal justice system and reinforcing their identity as a law-breaker, and therefore places the emphasis on using the minimum amount of intervention. Although perceptions may have changed somewhat between the earlier study and the subsequent evaluation of fast tracking, in order to evaluate fast tracking procedures one has to be aware of the practitioner perspectives (amongst other things), and how they form the basis for working relationships in fast tracking young offenders, since these perspectives have implications for the implementation of a programme. Thus, process research is partly about exploring the way in which those who implement programmes seek to realize their theoretical understanding of a phenomenon.

Conclusion

If such an approach to evaluation is adopted then there is likely to be a greater understanding of the structure and content of an initiative, the forces that have shaped it, the dynamics of a change situation, and of who is in control of it. This in turn makes it easier to address the all-important question of whether what happens in one location or set of circumstances might work elsewhere and in another set of circumstances. Evaluation involves much more than conducting a narrowly defined technical inquiry into whether or not something works. Properly conceived and carried through, evaluation is one of our best opportunities to study society and develop and refine theories about it. Too often, however, it is neither conceived nor executed in this way, the submission of the final report marking the end of the research, when it should be only a step in the process of inquiry.

Suggested readings

Cook, T.D. and Campbell, D.T. (1979) *Quasi-Experimentation*. Chicago: Rand McNally.
Everitt, A. and Hardiker, P. (1996) *Evaluating for Good Practice*. Basingstoke: Macmillan.
Pawson, R. and Tilley, N. (1997) *Realistic Evaluation*. London: Sage.
In addition the journal *Evaluation* contains articles on methodological issues regarding the evaluation of programmes.

Notes

I am grateful to Dr Alan France for comments on an earlier draft of this chapter, but what is said in it are my views alone and are not a commitment to a specific approach to the evaluation of the Communities That Care initiative which is referred to in this chapter.

1. This was attempted in a study which a colleague and I undertook during the 1980s. See Crow and Simon (1987) and Crow et al. (1989: Chapters 2 and 3).

2. Dr Alan France.

References

Agar, M. (1986) *Speaking of Ethnography*. London: Sage.
Bennett, T. (1996) 'What's new in evaluation research? A note on the Pawson and Tilley article', *British Journal of Criminology*, 36 (4): 567–573.
Bryman, A. and Cramer, D. (1990) *Quantitative Data Analysis for Social Scientists*. London: Routledge.
Cook, T.D. and Campbell, D.T. (1979) *Quasi-Experimentation*. Chicago: Rand McNally.
Cooper, H.M. (1989) *Integrating Research: a Guide for Literature Reviews*. London: Sage.
Crow, I. (1996) *Approaches to Youth Crime: a Study of the Views of Magistrates, Justices'*

Clerks and Social Workers. Centre for Criminological and Legal Research, University of Sheffield.

Crow, I., Cavadino, M., Dignan, J., Johnston, V. and Walker, M. (1996) *Changing Criminal Justice: The Impact of the Criminal Justice Act 1991 in Four Areas of the North of England.* Sheffield: Centre for Criminological and Legal Research, University of Sheffield.

Crow, I. and Simon, F. (1987) *Unemployment and Magistrates' Courts.* London: NACRO.

Crow, I., Pease, K. and Hillary, M. (1980) *The Manchester and Wiltshire Multifacility Schemes.* London: NACRO.

Crow, I., Richardson, P., Riddington, C. and Simon, F. (1989) *Unemployment, Crime and Offenders.* London: Routledge.

Everitt, A. and Hardiker, P. (1996) *Evaluating for Good Practice.* Basingstoke: Macmillan.

Farrington, D.P. (1997) 'Evaluating a community crime prevention program', *Evaluation*, 3 (2): 157–173.

Farrington, D.P. (1998) 'Evaluating "Communities That Care": realistic scientific considerations', *Evaluation*, 4 (2): 204–210.

Glaser, B.G. and Strauss, A.L. (1967) *The Discovery of Grounded Theory: Strategies for Qualitative Research.* New York: Aldine.

Kuhn, T.S. (1970) *The Structure of Scientific Revolutions.* Chicago and London: University of Chicago Press.

Nichols, G. and Taylor, P. (1996) *West Yorkshire Sports Counselling: Final Evaluation Report.* Sheffield: University of Sheffield, Leisure Management Unit.

Pawson, R. and Tilley, N. (1994) 'What works in evaluation research?', *British Journal of Criminology*, 34 (3): 291–306.

Pawson, R. and Tilley, N. (1996) 'What's crucial in evaluation research: a reply to Bennett', *British Journal of Criminology*, 36 (4): 574–578.

Pawson, R. and Tilley, N. (1997) *Realistic Evaluation.* London: Sage.

Pawson, R. and Tilley, N. (1998a) 'Caring communities, paradigm polemics, design debates', *Evaluation*, 4 (1): 73–90.

Pawson, R. and Tilley, N. (1998b) 'Cook book methods and disastrous recipes: a rejoinder to Farrington', *Evaluation*, 4 (2): 211–213.

Punch, K.F. (1998) *Introduction to Social Research: Quantitative and Qualitative Approaches.* London: Sage.

Rose, G. (1982) *Deciphering Sociological Research.* London: Macmillan.

7

REHABILITATION, RECIDIVISM AND REALISM: EVALUATING VIOLENCE REDUCTION PROGRAMMES IN PRISON

Roger Matthews and John Pitts

Contents

There has been growing interest in recent years in developing more appropriate and effective responses to those convicted of violent crime. A politically popular and increasingly widespread reaction on both sides of the Atlantic has been to 'get tough' and to 'crack down' on violent offenders whenever possible. The consequence of this approach has been to increase the level of punitiveness, to encourage the development of more austere and authoritarian prison regimes and to lengthen the sentences of those convicted of violence.

In relation to imprisonment the introduction and experimentation with 'boot camps' is symptomatic of the changing penal climate in both America and Britain. There is a growing emphasis on tough, military-style regimes which aim to instil discipline through vigorous training and exercise. Through the use of 'shock incarceration' young offenders, it is maintained, will be straightened out and will learn respect for authority. This approach, however, is far from new. In the late 1970s when the 'Iron Lady', Margaret Thatcher, came to power in the UK, the government announced the extension of 'short 'sharp shock' regimes to all of the existing detention centres. Based on a military model of correction and often employing ex-military personnel, these institutions aimed to impose discipline through strict regimes which emphasized regimentation, drill, training and intensive physical exercise. The idea was to instil discipline by giving 'slackers' extended periods of exercise and paramilitary forms of training. The problem was that the majority of the young men enjoyed it. This approach fitted well with their self-conceptions of masculinity, and for the majority engaging in paramilitary training was seen as more of a perk than a punishment (Shaw, 1985).

The result of this brief 'experiment' was that these juvenile institutions found it difficult to meet the prisoners' demands for exercise and training and eventually began to offer these activities not as a form of punishment, but as a *reward* for good behaviour. The net outcome of these regimes was to produce a group of very fit and strong young men with an inflated macho self-image whose affinity with aggression and interpersonal combat was enhanced. In effect, the specific deterrent value of these institutions was very low for many prisoners and, not surprisingly, the rate of readmission was fairly high.

The clear lesson which emerges from an examination of 'shock incarceration' is that such regimes are unlikely to address the issue of male violence but will tend to support and indeed encourage aggressive behaviour (McGuire and Priestly, 1995; Morash and Ricker, 1990; Sechrest, 1989). It should be noted, however, that in comparison to the apparent lawlessness of some of the existing juvenile institutions, in the UK at least, young men find the strict organization and discipline of these military-style institutions preferable. Recent reports by Her Majesty's Chief Inspector of Prisons on a number of young offenders' institutions, including Feltham in London and Glen Parva in Leicester, provide disturbing accounts of young people being left in their cells for most of the day or being exposed to continual bullying and intimidation by other prisoners. These institutions generate a milieu in which victimization, depression, and self-mutilation are all too common (HM Inspectorate of Prisons, 1996, 1997; Howard League, 1995).

To some extent the impetus to develop forms of 'shock incarceration' and to introduce 'boot camps' has been a function of the growing critiques of rehabilitation which circulated during the 1970s and early 1980s. The contention which was promulgated by an alliance of academics and policy-makers that rehabilitative programmes did not work encouraged a greater emphasis upon specific deterrence, which suggested that institutions should be as unpleasant and uncomfortable as possible, and on the incapacitation and warehousing of offenders in order to take them off the streets (Zimring and Hawkins, 1995). However, it became all too apparent during the 1980s that simply warehousing offenders in poor conditions was not a position of neutrality, since when young offenders are placed in custody for a period of time they do not emerge from these institutions the same as they went in. That is, it became increasingly obvious to the majority of those who worked in or closely with penal institutions that leaving prisoners locked up in their cells for most of the day, or simply engaging in a policy of containment, was a recipe for producing more disturbed, more violent, more committed and more marginalized offenders who, through their exposure to these often lawless regimes, tended to become either more aggressive and defiant, or withdrawn and depressed (Little, 1990).

Interestingly, it was particularly amongst those groups who had been seen as being 'beyond salvation' – sex offenders and violent offenders – that it became more evident that rehabilitation programmes needed to be introduced within penal institutions and it was in relation to these groups of offenders that a range of promising and imaginative interventions have been introduced (Sampson, 1994). These programmes include forms of behavioural therapy, group therapy and counselling and have been developing steadily in different institutions alongside existing educational and training programmes (Palmer, 1992).

The implementation of programmes in the UK has, however, been very patchy. Although there have been frequent reports about the benefits of specific programmes in different institutions, much of this work is relatively new and the level of evaluation has been poor. The expansion of different types of rehabilitative programme on both sides of the Atlantic over the past decade or so has raised the issue of 'what works' and has brought to the fore the question of evaluation. Following the retreat of Martinson and his colleagues from the 'nothing works' position which was so widely quoted in the 1970s, it has gradually become recognized that some programmes appear to work with some people (Martinson, 1974, 1979). The problem remains to clearly identify which programmes work and for whom under what conditions (Gendreau and Ross, 1987).

One approach which has gained a considerable amount of attention and has been referred to as 'the state of the art in correctional intervention' (Robinson, 1995), and which has been subject to extensive evaluation, is the Cognitive Skills Programme. This programme, which was developed in Canada in the late 1980s, is designed to replace 'faulty' thinking by 'straight' thinking and operates on the assumption that 'faulty' thinking in the form of limited cognitive skills and reasoning capacity produces crimogenic traits in individuals which can be effectively addressed through a course of instruction lasting normally 35 hours over an 8–12-week period (Porporino and Robinson, 1995). In Canada over 5,000 prisoners have been through the Cognitive Skills Programme and the researchers claim considerable success in reducing the rate of recidivism and reconviction amongst completers. On the basis of the reported effects of the programme, the Prisons Department in England and Wales has imported this approach wholesale and in 1998 over 60 institutions in the UK ran programmes of this type. Although there has been little in the way of systematic evaluation of these programmes in England and Wales to date, the considerable investment which the Prisons Department has made suggests that a critical review would be prudent.

The Cognitive Skills Programme

The development of the Cognitive Skills Programme by the Canadian Correctional Service was based on the original work of Ross and Fabiano (1985).

These programmes focus on changing the thinking patterns which are seen to typify the ways in which many offenders attempt to solve problems and to make decisions. In essence, the programmes are designed to improve reasoning by overcoming rigidity and the perceived lack of problem-solving skills amongst offenders. The overall aim is to improve self-control, negotiation skills and the management of emotions. The emphasis in the programme is not so much on what the offenders think but rather on *how* they think. The training sessions involve a combination of games, puzzles, reasoning exercises and discussion which are designed to build up reasoning skills in a cumulative way (Blud, 1996).

Selection for the programme is normally arranged through case management officers and candidates are assessed by programme delivery staff in order to ensure that they possess the type of cognitive deficit addressed by the programme and are suitably motivated to participate. After eligible candidates have been identified, they are randomly assigned to receive the programme immediately or to be placed on a waiting list for the next available programme. Those who remain on the waiting list without actually receiving the programme are used as a control group which can be used as the basis for post-release outcome comparisons.

Prospective candidates are assessed in relation to needs and risk factors. By needs the researchers refer to what are seen as those cognitive deficits which are believed to be associated with criminal behaviour such as pro-criminal attitudes, substance abuse and poor reasoning skills. A risk scale is also employed which is based on criminal history and candidates are divided into low and high need offenders and low and high risk offenders for the purpose of analysis and in order to identify the groups, if any, with which the Cognitive Skills Programme is most effective.

The problem of evaluation

The evaluation of prison programmes is beset with difficulties. A major problem is maintaining programme integrity: that is, ensuring that the programme is implemented in a consistent manner and in a way which follows the original design. The Canadian researchers appear to be sensitive to this problem and have gone to considerable lengths to ensure that the programme is implemented in the required manner by training delivery staff and by regularly monitoring its development. There is also an important issue concerning the appropriate 'dosage' given to each person, since it might be expected that different types of prisoner with different degrees of 'cognitive deficiency' would require input of different lengths and intensities. The Cognitive Skills Programme runs a standard programme, although there has been some discussion about the possible advantages of running more intensive programmes for more serious and committed offenders. There are, however, two other problems which tend to arise in relation to the evaluation of rehabilitative programmes. The first is the development of an adequate methodology and the second involves devising appropriate measures of success.

In relation to methodology reference has already been made to the creation of control groups which are designed to provide a benchmark against which changes in the treatment group can be assessed. This has become a well-established strategy in criminological and penological research, but as Pawson and Tilley (1997) have pointed out, there are serious questions which arise in relation to the appropriateness of quasi-experimental methods to the analysis of voluntary rehabilitative strategies in prisons. This is principally because the value and reliability of the quasi-experimental method is dependent on the use of randomized samples whereas the participants in rehabilitative programmes tend to involve largely self-selecting groups. The other major limitation of the quasi-experimental method is that it says little about the reasons for the transformation of those who participate in the programme and instead assumes that any difference in behavioural outcomes is a direct result of the application of the initiative. In this scenario the treatment is seen to work 'on' prisoners although the exact mechanisms through which change has occurred are not clear. This is the so-called 'black box' problem in which the process of transformation is not explained but simply inferred from the outcome.

A critical element in achieving some form of behavioural change, Pawson and Tilley (1997) argue, is the motivation and predisposition of the offender. In opposition to the contention that programmes work 'on' individuals, they argue that programmes actually work 'through' individuals since without the co-operation and commitment of the subjects even the best-designed initiatives are likely to be ineffective. In short, it is not that rehabilitative programmes work *per se* but that they work through interaction with those subjects who choose to make them work. Thus, in opposition to the question of 'what works' we should be asking, how do subjects who participate in different programmes mobilize the available opportunities to achieve certain outcomes? Thus:

> Social programmes involve a continual round of interactions and opportunities and decisions. Regardless of whether they are born of inspiration or ignorance, the subject's choice at each of these junctures will frame the extent and nature of change. What we are describing here is not just the moment when the subject signs up to enter a programme but the entire learning process. The act of volunteering merely marks a moment in a whole evolving pattern of choice. Potential subjects will consider a programme (or not), cooperate closely (or not), stay the course (or not), learn lessons (or not), apply the lessons (or not). Each one of these decisions will be internally complex and take its meaning according to the chooser's circumstances. (Pawson and Tilley, 1997: 38)

Individuals will volunteer for programmes for a number of different reasons. They may genuinely desire to change, they may simply be bored or they may see it as a way of gaining early release or increasing their eligibility for parole. The nature of the offenders' motivation therefore appears to be the key variable and evaluation of corrections needs to develop a more comprehensive and less mechanistic analysis of the ways in which different subjects make

choices and respond to the range of opportunities offered by different programmes. The process of evaluation, therefore, needs to make direct reference
to the make-up and motivations of the volunteers rather than concentrate on
the comparison between the treatment and the control group.

In a similar vein Elliott Currie (1986) has suggested it might be more
profitable to think in terms of 'opportunity models' rather than 'deficit
models' when designing interventions. That is, rather than developing pre-
dominantly reactive compensatory strategies which concentrate on 'repairing'
the damage, interventions can be seen as useful vehicles for providing opportunities for particular user groups to change and develop.

Evaluating cognitive skills programmes

In presenting the results of cognitive skills programmes the evaluators
provide a range of findings which claim that in general those who completed
the course have a lesser rate of reconviction and readmission than other
categories of offender. Overall, however, the results are not particularly
impressive since 19.7 per cent of programme completers were reconvicted in
a 12-month period compared to 24.8 per cent of the control group (see Figure
7.1). The most interesting finding shown in Figure 7.1 is the rate of reconviction and readmissions amongst the 'dropouts' who did not complete the
programme, either because of their refusal to participate or because of their
dismissal by programme delivery staff. In the above cohort some 14 per cent
of programme participants dropped out during the course. By definition the
majority of these dropouts lacked motivation and co-operation and not
altogether surprisingly had a much higher level of recidivism than all other
types of participant.

By removing the dropouts from the calculations of the programme and by
only comparing the completers with the control group, a false point of
comparison is constructed. Clearly, if the control group had engaged in the
programme a predictable percentage would have dropped out and this group
would be expected to have a high level of recidivism. Thus if the sub-group of
potential dropouts is removed from the control group the rate of recidivism
for this group would decrease significantly. In fact, if we were to adjust the
figures for the control group taking the probable effect of dropouts into
consideration the net difference in recidivism between the treatment group
and the control group would be minimal.

One of the major reasons why the Cognitive Skills Programme might be
less effective than many of its proponents have suggested is because there is a
serious conceptual flaw which runs through the heart of this approach.
Because the programme aims to address principally *how* people think rather
than *what* they think and because the aim is to produce more rational and
coherent thinkers, it is possible that it will produce more rational and
accomplished offenders. By the same token, these courses could produce more
adept criminals who are better able to avoid being caught and convicted. The
conceptual slippage which underpins these programmes is an erroneous

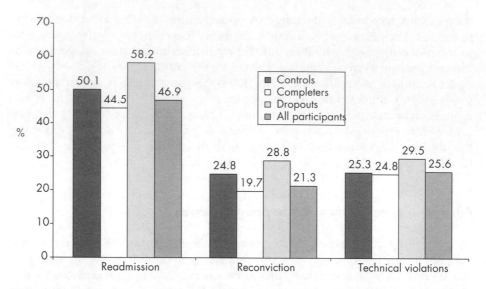

Figure 7.1   *Outcomes by programme participation. Source: Robinson, 1995.*

equation of 'logical thinking' with being 'law abiding' and 'straight thinking' with 'going straight'. The assumption is that many people engage in crime because they lack the reasoning power and the problem-solving skills to deal with situations in law-abiding ways. The reality, of course, is that for many offenders who resort to violence or deception these strategies are actually quite effective in realizing certain objectives.

The advocates of the programme may respond to this criticism by pointing out that although the overall differences may be slight, the programme appears to be successful with certain types of offender. Robinson (1995) for example claims that the Cognitive Skills Programme is most successful with low risk violent and sex offenders, and particularly amongst the 25–39 age group. The problem, however, is that he and his colleagues provide no real causal explanation of why the programme might work better with some types of offender rather than others and therefore leaves the issue open to speculation.

Much of the validity of the analysis presented by the evaluators of the Cognitive Skills Programme is dependent upon the appropriateness of the measures of success which are employed. Two measures are used. The main one is the level of recidivism, which is measured in terms of reconviction and readmission rates, while the other measure is the number of offenders in the treatment group and control group who are given early release or parole.

The second measure is particularly suspect since it is fairly transparent even to those who have 'faulty' cognitive skills that simply by volunteering for programmes of this type they are likely to increase their chances of parole, unless of course they drop out. As a measure of success, therefore, the offer of early release or parole does not provide a very good indicator of the programme's success. A more substantive and widely used criterion for success is the level of recidivism. The obvious attraction of using this indicator derives from the fact that gathering reconviction data and recidivism data is relatively straightforward. However, as a criterion for evaluating rehabilitative programmes within prisons it is a weak measure. Since recidivism is employed as the main point of reference in over 90 per cent of evaluations for assessing the achievements of different programmes it is necessary to critically examine its utility and its appropriateness for this task.

The problem of recidivism as a measure of success

One of the lessons which should have been learned from the debate involving Robert Martinson (1974) on 'What works?' is that recidivism is a very dubious measure of programme effectiveness. The danger is not only that the causal links between the programme and the point of measurement remain hidden and largely forgotten, but that recidivism itself is nothing like as robust a measure as many of its advocates claim and that it can be formulated and defined in a number of different and even conflicting ways. The critique of Martinson by Ted Palmer (1975) and others made the important point that Martinson's own data did not warrant the conclusion that nothing works. On the contrary Martinson's research suggested that some programmes worked with some participants in certain settings. Although Palmer's call for specificity was undoubtedly correct, he remained within the same discursive formation as Martinson in that he continued to use recidivism as the primary point of reference when assessing the effectiveness of different programmes. This was unfortunate because the limitations of Martinson's original analysis stemmed not only from his tendency to over-generalize but also from his uncritical use of recidivism as the main means of evaluating programmes. It was apparent that many of the programmes which he examined had a number of positive effects although they may have had little influence on the probability of offenders being rearrested or reconvicted (Matthews, 1987).

Recidivism is an unstable measure which is calculated by different researchers in different ways. In general it is used to refer to one or a combination of four different measures – rearrest rate, reconviction rate, readmission rate and reoffending rate. Although the rearrest rate is often used as a convenient point of reference, it is widely recognized that the frequency with which offenders are likely to be arrested after leaving prison may be more a function of the police's preoccupation with rounding up 'known suspects' than it is to do with the actual level of offending. In order to remove those cases which involve trivial or spurious incidents some researchers prefer to use reconviction data since this tends to screen out a percentage of minor incidents. However, the dilemma arises that a percentage of reoffenders may avoid conviction through technicalities or for contingent reasons but they are still identified as examples of 'success'. Readmission rates are preferred by some researchers since they appear to present much 'harder' data because they suggest that the offence must have been relatively serious in nature if it has resulted in the offender being sent back to prison (Jones, 1996). Although this is likely it may be that the offence was significantly *less serious* than the offence for which the person was originally imprisoned. In these cases what might be viewed as a failure from one vantage point could be seen as a success from another.

Since none of these three measures offers a very accurate indicator of the actual level of reoffending, some researchers prefer to ask ex-prisoners themselves whether they have reoffended or not in a given period of time. Although this approach promises to reveal important information about the nature and the seriousness of offences committed after release from prison, the ex-prisoner has an enormous disincentive to reveal details of offences which he or she may have committed since leaving prison, if they have not been arrested or convicted for these offences. Self-report reoffending studies are likely to grossly underestimate the true level of reoffending and provide a limited, if not distorted, picture of the patterns of post-release offending. In those exceptional situations in which trust is established between ex-prisoners and researchers and confidentiality is assured it may well be that high levels of reoffending are recorded, although there have been few arrests or reconvictions. This very real possibility reveals the weakness of recidivist measures which are based purely on rearrest and reconviction scores as measures of programme effectiveness.

There are other major problems with using recidivism as a measure of success. These are associated with the often arbitrary period of time – normally one or two years – which is taken as the relevant period over which to examine reconviction or readmission (Maltz, 1984). Clearly, the impact of any programme is likely to diminish over time and to use the fact that someone has been reconvicted within a two-year period as a criterion for the failure of a particular programme is unreasonable. Further, the use of recidivism suffers from the disadvantage that even if ex-prisoners are convicted of less serious offences, or even if they are committing offences significantly less frequently than before they entered prison, they may still be labelled as recidivists and held up as examples of programme failure (Wilson, 1985).

In short, recidivism is an unreliable measure of programme effectiveness and, although the public and policy-makers may have an interest in evaluating programmes' impact on reoffending, recidivism in its various guises rarely presents an accurate picture and is often too far removed from the specific intervention itself to have direct relevance. By the same token, it is widely recognized that the pressures to reoffend after leaving prison may be a function more of the level of interpersonal and familial support, the availability of suitable accommodation and the prospect of employment than of the participation in rehabilitative programmes in prison, no matter how well thought out they may be.

So if recidivism, however calculated, is employed as a measure of success, it should at best serve as an ancillary or back-up measure rather than as the main point of reference for evaluation. Too many promising initiatives have been sacrificed on the altar of recidivism and it is necessary, if these programmes are to be systematically developed and evaluated, for more realistic measures to be found.

Developing realistic intermediate measures of success

It seems a reasonable starting point to attempt to measure rehabilitative programmes in terms of what they can control rather than in relation to outcomes over which they have little or no influence. Moreover, it is essential to determine what it is about certain programmes which makes the transformation of different individuals possible.

In relation to the measurement of changes in behaviour and attitudes of those engaged in different programmes there is also a need to be realistic. The possibility of reversing 15 or 20 years of socialization within an 8- or 12-week programme is myopic, to say the least. Programmes which claim, or indeed even aim, to achieve such outcomes in such a short period should be treated with extreme caution. The majority of offenders are actually in prison for relatively short periods of time and a significant percentage are held on remand. At the same time it is overly optimistic to believe that whether individuals engage in rehabilitative programmes or not that they plan to stop offending when they leave prison. Thus any assessment of the effectiveness of specific programmes to reduce or stop offending at least needs to take into account the criminal aspirations of prisoners at the point when they leave prison.

In short, it is necessary to move away from a zero-sum conception of rehabilitation and from the notion that the aim of rehabilitative programmes is to turn bad people into good people or committed criminals into law-abiding citizens. The aims of rehabilitative programmes must be more diverse and more modest. They need to be designed to achieve a number of different objectives at a number of levels, since even gains at the margin are gains.

If we move away from the notion that programmes either do or do not work and instead ask what it is about a particular programme which makes change possible, we can develop a range of criteria by which to measure success. Courses may offer new skills, training, question values and preferences, or increase self-awareness and self-esteem. Apart from programme completion rates and ongoing assessments of achievements of those involved in various courses, forms of evaluation could involve a number of intermediate measures, such as changes in anti-social attitudes, developing the conditions for independent living, and developing literacy and numeracy skills. More specifically, in relation to violence, the forms of evaluation which could be adopted might include the number of violent incidents in prison, self-report surveys of aggressive behaviour in prison as well as staff evaluations.

Such criteria are already used in various ways in some forms of evaluation but often have an ancillary or back-up role. These measures need to be developed and refined, but the key point is to develop a range of measures which can accommodate small scale but significant changes of attitude and behaviour and which can act as a useful means of beginning to explain why certain programmes work with certain types of offender.

Explaining why programmes work

Understanding something about the processes through which certain programmes work with certain types of offender is critical to the task of evaluation. A programme which achieves a desired outcome may do so because it encourages self-realization, increases personal skills or awareness, or promotes social acceptability because it allows the inmate to function in a wider range of settings (Pawson and Tilley, 1997). The evaluator should aim to find out which of these, or other processes, are in operation. In order to achieve this objective it is not enough to rely on purely quantitative data which concentrates on correlations and the patterning of variables; evaluation also needs to include more qualitative and 'intensive' data gained from discussion with those who have actually participated in the programme (Sayer, 1992). Detailed interviews with participants are necessary to find out what it was about the intervention that is most likely to encourage specific changes.

The overall aim is to develop causal explanations of the processes by which rehabilitation programmes make certain forms of change possible. By combining 'extensive' research which looks at regularities and common patterns with 'intensive' forms of investigation which address the question of how the process works in an individual case or a small number of cases, it is possible to identify causal processes within broader patterns of difference. As Van Voorhis et al. have observed:

> Most of what we have learned has come from small, well controlled studies, rather than from large scale initiatives. We often neglect to replicate the most successful programs, and we often fail to create a clear prototype that will facilitate replication. The paucity of evaluation research, and the failure to replicate successful strategies embodies the tragedy of many social programming endeavors. Most of what we know to be effective is not currently in practice, most of the seemingly good ideas, which constitute current practice, have not been or are not being tested. (Van Voorhis et al., 1998: 2)

The possibility of an offender benefiting from a particular intervention may be to do with his or her background and biography. These developments may well go far beyond the types of criminal histories which are employed in calculating 'needs assessments' and may include a range of social and structural factors which have predisposed certain types of people to engage in crime. Thus the biographical conditions which have led to the individual's involvement in crime require some theorization if meaningful categories of 'crimogenic needs' are to be developed.

The available evidence from arrest data and victimization surveys shows that violent offenders are not randomly distributed throughout the population. Instead, it indicates that juvenile violence is primarily an urban phenomenon and is associated with poverty and generally weakened social institutions, such as schools, households and neighbourhood groups (Fagan, 1990). An effective strategy for dealing with youth violence needs to take these factors into account.

One critical element which is conspicuously absent from the evaluation studies on the Cognitive Skills and other similar programmes is the influence of the general prison culture. Differences in prison regime and styles of prison management can have a considerable influence on the climate in which programmes are carried out and experienced. The inmate subculture in prison, as penologists have consistently pointed out, can profoundly influence self-esteem and the prisoners' sense of personal safety. A recent study on victimization in prisons in England and Wales found that just under half of all young offenders had been assaulted, robbed or threatened with physical violence during the previous *month* (O'Donnell and Edgar, 1996). Attempting to develop and evaluate violence reduction programmes in those prisons which are riddled with high levels of interpersonal violence, bullying and intimidation is always going to be difficult. In contrast, a prison like Grendon, in southern England, which operates as a 'therapeutic community' is able to conduct courses and programmes within an overall framework in which violence and drug-taking in the prison as a whole are not tolerated and consequently can increase the chances of success of the various programmes which it runs for sex and violent offenders (Genders and Player, 1995).

Conclusion

Violence has been identified as a major problem in recent years, and various legal and extra-legal strategies have been developed to address this issue. Of offenders who come to court, a significant percentage will end up in prison. The longer sentences that are being handed out to those convicted of violence means that violent offenders are forming a larger proportion of the prison population and this development, in turn, is transforming prisons into potentially more volatile institutions. It is becoming evident, however, that 'get tough' militaristic approaches do little to address the question of violence and are more likely to reinforce the offender's predisposition towards aggressive behaviour. Yet lawless and disorganized regimes in which young prisoners are ritually bullied and/or locked in their cells for much of the day are in many respects worse. Doing nothing in prisons and leaving prisoners to their own devices is a recipe for disaster. Prison itself is a debilitating and stigmatizing experience. If the aim is to reintegrate offenders into society and reduce the level of violence, constructive and imaginative interventions need to be developed.

During the 1980s a number of promising initiatives were introduced in prisons on both sides of the Atlantic. Many were patchy and short-lived, and for the most part they were not subject to systematic evaluation. The Cognitive Skills Programme, which was developed in Canada, has been held up as a good example of a well-designed, effective and properly evaluated programme. The achievements claimed on its behalf have persuaded the Prisons Department in England and Wales to adopt this general approach and to make it a central part of their rehabilitative strategy. A critical examination of the programme and its evaluation reveals, however, that there are serious flaws not only in its conceptualization but also in the way it has been operationalized. The major failing of the evaluation is that it tells us very little about *what* works and even less about *how* it works. There is no real examination of the causal processes at work and the main criterion for success – recidivism – remains an imprecise and uncertain measure.

Rather than continue to evaluate programmes principally in terms of recidivism, as occurs in the vast majority of cases, it has been suggested that it is necessary to develop a range of realistic intermediate measures which combine quantitative and qualitative data through the use of 'extensive' and 'intensive' forms of research. In this way it is possible to identify causal mechanisms and to examine their operation within different contexts.

Finally, it is probably about time that we stopped repeatedly asking ourselves the question 'What works?' as if the solution to the problem of achieving rehabilitation in prison is just a matter of finding the right instrument of reform. Posing the question in this mechanistic way can too easily lead us to lose sight of the interactive nature of personal change and development and the key role which prisoner co-operation plays in determining outcomes.

Suggested readings

Martinson, R. (1974) 'What works? Questions and answers about prison reform', *Public Interest*, 32: 22–45.
McGuire, J. (ed.) (1995) *What Works: Reducing Re-offending*. London: Wiley.
Pawson, R. and Tilley, N. (1997) *Realistic Evaluation*. London: Sage.
Sayer, A. (1992) *Method in Social Science: A Realist Approach*. London: Routledge.

References

Blud, L. (1996) 'Cognitive skills programmes in prison', in S. Hayman (ed.) *What Works With Young Prisoners?* London: ISTD. pp. 22–28.
Currie, E. (1986) *Confronting Crime: An American Challenge*. New York: Pantheon.
Fagan, F. (1990) 'Treatment and reintegration of violent offenders: experimental results', *Justice Quarterly*, 7 (2): 233–262.
Genders, E. and Player, E. (1995) *Grendon: A Study of a Therapeutic Community*. Oxford: Clarendon Press.
Gendreau, P. and Ross, R. (1987) 'Revivification of rehabilitation: evidence from the 1980s', *Justice Quarterly*, 4 (3): 349–407.
HM Inspectorate of Prisons for England and Wales (1996) *Feltham: Report of a Full Inspection*. London: Home Office.
HM Inspectorate of Prisons For England and Wales (1997) *Young Prisoners: A Thematic Review*. London: Home Office.
The Howard League (1995) *Banged Up, Beaten Up and Cutting Up: Report of The Howard League Commission of Inquiry into Violence in Penal Institutions for Teenagers under 18.* London: The Howard League.

Jones, P. (1996) 'Risk prediction in criminal justice', in A. Harland (ed.) *Choosing Correctional Options That Work*. London: Sage. pp. 33–69.

Little, M. (1990) *Young Men in Prison*. Aldershot: Dartmouth.

Maltz, M. (1984) *Recidivism*. Orlando, FL: Academic Press.

Martinson, R. (1974) 'What works? Questions and answers about prison reform', *Public Interest*, 32: 22–45.

Martinson, R. (1979) 'New findings, new views: a note of caution regarding sentencing reform', *Hofstra Law Review*, 7: 243–258.

Matthews, R. (1987) 'Decarceration and social control: fantasies and realities', in J. Lowman et al. (eds) *Transcarceration: Essays in the Sociology of Social Control*. Aldershot: Gower.

McGuire, J. and Priestly, P. (1995) 'Reviewing what works: past, present and future', in J. McGuire (ed.) *What Works: Reducing Reoffending*. London: Wiley. pp. 33–35.

Morash, M. and Ricker, L. (1990) 'A critical look at the idea of boot camps as correctional reform', *Crime and Delinquency*, 36 (2): 204–222.

O'Donnell, I. and Edgar, K. (1996) *Victimisation in Prisons*. Research Findings No. 37. London: Home Office Research and Statistics Directorate.

Palmer, T. (1975) 'Martinson revisited', *Journal of Research in Crime and Delinquency*, 12: 133–152.

Palmer, T. (1992) *The Re-Emergence of Correctional Intervention*. London: Sage.

Pawson, R. and Tilley, N. (1997) *Realistic Evaluation*. London: Sage.

Porporino, T. and Robinson, D. (1995) 'An evaluation of the reasoning and rehabilitation program with Canadian federal offenders', in R. Ross and R. Ross (eds) *Thinking Straight*. Ottawa: Air Training Publications. pp. 29–46.

Robinson, D. (1995) *The Impact of Cognitive Skills Training on Post Release Recidivism Among Canadian Federal Prisoners*. Ottawa: Ottawa Correctional Services.

Ross, R. and Fabiano, E. (1985) *Time to Think: A Cognitive Model of Delinquency Prevention and Offender Rehabilitation*. Johnson City, TN: Institute of Social Science and the Arts.

Sampson, A. (1994) *Acts of Abuse: Sex Offenders and the Criminal Justice System*. London: Routledge.

Sayer, A. (1992) *Method in Social Science: A Realist Approach*. London: Routledge.

Sechrest, D. (1989) 'Prison boot camps do not measure up', *Federal Probation*, 63 (3): 18–20.

Shaw, S. (1985) 'Reflections on "short, sharp shock"', *Youth and Policy*, 13 (Summer): 1–15.

Van Voorhis, P., Cullen, F. and Applegate, B. (1995) 'Evaluating interventions with violent offenders: a guide to practitioners and policy makers', *Federal Probation*, 59 (June): 17–28.

Van Voorhis, P., Cullen, F. and Applegate, B. (1998) 'Evaluating interventions with violent offenders: a guide to practitioners and policy makers'. Mimeo, University of Cincinnati.

Wilson, J. (1985) *Thinking about Crime*, 2nd edition. New York: Vintage.

Zimring, F. and Hawkins, G. (1995) *Incapacitation: Penal Confinement and the Restraint of Crime*. New York: Oxford University Press.

8

CRIME SURVEYS AND THE MEASUREMENT PROBLEM: FEAR OF CRIME

Jason Ditton, with Stephen Farrall, Jon Bannister and Elizabeth Gilchrist

Contents

If the 1980s were something of a golden era for fear-of-crime research, the late 1990s have seen the emergence of enough nagging doubts to justify a period of introspective reflection, followed perhaps by a rather different approach to the subject.

The many local crime and fear-of-crime surveys that have been conducted typically used questionnaires and asked questions in ways derived from the structure and methods of the British Crime Survey (BCS). The BCS is run every two years with a freshly selected multi-staged, clustered, cross-sectional sample. It is not designed to test fear-reduction or crime-reduction measures. When instruments derived from it are used at the local level, irrespective of differing local circumstances, they all seemed to generate data indicating a broadly similar picture of the fear of crime. Everywhere, women and the elderly were found to be more fearful.

Worse still, these levels of fear seemed to be stubbornly immune to any reduction fuelled by local environmental and other improvements. Locally measured levels of crime itself, on the other hand, did go up and down, but here the problem was a progressive inability genuinely to believe that this was an effect of local environmental improvements, and was not just a local area benefiting from a national crime rate downturn, or, even more depressingly, from some random fluctuation in local rates of offending.

At a slightly deeper level, some analysts began to wonder what the 'fear of crime' actually was? What does it mean to say, for example, that '60 per cent of those questioned were very worried or a bit worried about being burgled'? Is 'worry' the same as 'afraid'? Are both 'fear'? It has even been asked whether or not there is anything wrong with people being fearful of becoming crime victims.

Against this background, we embarked in 1994 on a major study, funded by the Economic and Social Research Council, designed not as yet another crime *survey*, but instead as a study of crime survey *methodology*. This involved first conducting a large number of qualitative interviews to determine the full range of responses to the prospect of or reality of crime victimization; and what precisely, for different people, these actually meant. Focus groups were then assembled to discuss these issues, and finally we condensed what we had been told into an extensive questionnaire which included both the conventional and our 'new' questions. This questionnaire was eventually administered to a huge, pure random sample of over 1,600 Scottish adults. Some results have been published already (Ditton et al., 1998, 1999a, 1999b, 1999c; Ditton and Short, 1999; Farrall et al., 1997a, 1997b; Gilchrist et al., 1998), others are currently in press (Ditton; Ditton and Farrall; Farrall et al.) and still others are in various stages of dissemination (Farrall, et al.).

The need to measure fear

At the same time as basic questions are being asked about fear-of-crime research methodology – something which suggests that a period of reflection and contemplation might be in order – the Crime and Disorder Act 1998 has laid statutory duty upon every local authority to 'formulate and implement, for each relevant period, a strategy for the reduction of crime and disorder in the area'.

It is likely that any novel research underpinning such strategic planning will be based on a locally organized crime survey (although see Chapter 9 in this volume), and equally likely that questions will be asked of the fear of criminal victimization, in addition to those probing victimization itself. No doubt any thus formulated strategy will propose to reduce both. However, while it is difficult to think of any good reason why victimization should not be reduced, the fear of victimization is another matter entirely.

Take one deep conceptual issue. Although it goes against the common-sensical grain to say this, why should fear be reduced? Little if any justification is ever advanced in support of this common policy thread. Much *post hoc* muttering about its 'negative impact on the quality of life' is normally paraded in the place of calm reasoning, but 'fear' (or whatever it is that questions on the 'fear' of crime actually tap into) is a very basic drive whose retroductively supposed role is the protection of those who fear. As a result the whole of Britain is now being tasked to reduce something it does not understand and cannot measure reliably. It is arguable that it should not be reduced, although, perhaps perversely happily, attempts to reduce it usually fail.

Problems with current approaches

The briefest possible history of the 'fear' of crime would be this. There was no 'fear' of crime in Britain until it was discovered in 1982. Crime surveyors liked it because whereas only about five interviewees in 100 could recall a crime victimization from the recall period (usually the previous year), 100 out of 100 could give usable data about their 'fear' of crime. Politicians and policy-makers liked it because it seemed more amenable to manipulation and reduction than crime itself. Rates of 'fear' of crime, at one point, seemed about to become more important than rates of crime itself. People set about energetically trying to reduce the 'fear' of crime. They failed.

First of all, rates of 'fear' of crime seem remarkably stable. An analysis of national crime survey reports in the 1990s (the British Crime Survey reports for 1994 and 1996, and the Scottish Crime Survey reports for 1984, 1992 and 1996) makes it clear that:

1   The number of respondents who claim that they feel a bit or very unsafe when walking alone in their area at night is 35 per cent ± 6 per cent;
2   The number of respondents who claim that they feel a bit or very unsafe when home alone at night is consistently 10 per cent ± 1 per cent;
3   The number of respondents who claim that they are a bit worried or very worried about being burgled is 60 per cent ± 5 per cent;
4   The number of respondents who claim that they are a bit worried or very worried about being mugged is consistently 47 per cent ± 1 per cent;
5   The number of respondents who claim that they are concerned about being a crime victim is consistently 55 per cent ± 3 per cent.

This is over a period when police-recorded crime rates had been stabilizing and then falling in England and Wales, and falling rather more dramatically in Scotland. A criminological maxim appears to be emerging here: it seems that rates of 'fear' of crime may climb when the crime rate climbs, but fail to fall when the crime rate falls (Perring, quoted in Horne, 1996: 321). Two examples of the stubbornness of the 'fear' of crime serve to illustrate this. The first is from research conducted in Scotland, the second from one of the most sophisticated studies ever conducted in England and Wales.

The Scottish research is reported in full in Nair et al., 1993. To summarize, between two substantial crime and safety audits conducted in Glasgow's Castlemilk public housing estate, a small child was brutally murdered at dusk in a small park called, locally, the 'pond'. As it happened, the local housing department was about to substantially upgrade the tenement flat housing stock adjacent to the pond area, and this included wholesale flat refurbishment, remote door entry systems, increased door and window security, and so on. After the murder, the district council decided swiftly also to

Table 8.1 *Summary of effect of crime-fear prevention activity*

	SC no action	SC low intensity	SC med. intensity	SC high intensity	all SC action areas	comparison cities
% houses burgled 1+ times last year BEFORE	8.9	10.3	12.7	13.4	11.6	12.0
% houses burgled 1+ times last year AFTER	10.2	9.3	9.9	7.6	9.1	12.4
Before–after % change (real)	+15	–10	–22	–43	–21	+3
% worried + very worried about burglary BEFORE (1990)	67.5	71.4	72.5	69.4	69.2	71.2
% worried + very worried about burglary AFTER (1992)	69.3	69.1	73.7	74.1	70.3	68.6
Before–after % change (all)	+3	+3	+2	+7	+2	–4
Before–after % change (those aware)	0	+10	–6	–9	+1	–17
No. schemes	–	34	40	41	96	
Unweighted N	3,138	1,134	590	710	5,576	2,099
n (aware)	633	259	107	212	1,211	424

This table is constructed from Ekblom et al.'s (1996) data as given on pages xiv, 13 and 70. 'Low' intensity action is defined as an investment of less than £1 per household; 'medium' intensity between £1 and £13 per household, and 'high' as more than £13 per household.

SC = Safer Cities

improve the external pond area by enhancing lighting, widening and resurfacing paths, and cutting back trees, bushes and undergrowth.

A small number (33) of local residents were interviewed both before and after the improvements were carried out. They should have felt safer, but, to summarize the results, there was a 7 per cent net *increase* in feeling 'unsafe' when at home at night, a substantial *increase* in the use of precautionary measures (such as entry phones. Of course, these had been provided, but an increase in their use is normally seen as indicative of increased fear); a net *increase* in those who avoided going out alone at night; and only a 1 per cent increase in the number feeling 'safe' when out alone after dark. The published report concluded with the words, 'if massive improvements to domestic safety measures coupled with enhanced local street lighting, path widening, and so on fail to make a significant impact on residents' fear of crime, what is there left to try?'

The second example is from a massive study conducted by the Home Office (Ekblom et al., 1996). Their goal was to see if the Safer Cities investment in burglary prevention actually reduced both burglaries, and the fear of burglary. Their results are given in Table 8.1.

What does this tell us? First, and by looking at the data in row 3 of the table, it was clear that the more crime-preventative action taken, the greater the *decline* in the burglary rate. So, at least something works. But second, and

Table 8.2   *Pedestrians' expectations/realizations of the effects of improved street lighting*

	Will (before) N=425	Did (soon after) N=340	Did (much later) N=413
Will/did pedestrian traffic increase?	344 (83%)	129 (49%)	157 (44%)
Will/did number of unpleasant incidents decline?	296 (74%)	66 (28%)	71 (21%)
Will/did people's fear of crime decline?	282 (69%)	97 (35%)	91 (25%)
Will/did the amount of crime decline?	233 (59%)	49 (20%)	57 (17%)

this data is in row 6, there was an *increase* in the self-declared levels of worry about burglary, and this was greatest (at 7 per cent) in those areas which had implemented the most crime-preventative action. All is not lost, however. Row 7 is included as it shows that those actually aware of low intensity crime-preventative action became more worried, but those aware of high intensity became less so. So far, so good. However, those unaware but receiving low intensity crime-preventative action became less worried: those unaware but receiving high intensity crime-preventative action became more worried. At this point, one might legitimately ask: do people know what makes them safer?

Some research conducted in the early 1990s into the effects of improved street lighting on crime and the fear of crime in Glasgow suggests some answers. (The research is reported, *inter alia*, in: Ditton and McNair, 1994; Ditton and Nair, 1994; Ditton et al., 1993; Ditton et al., 1996; McNair and Ditton, 1995; Nair and Ditton, 1994; and Nair et al., 1997.) One of a number of different research measures adopted in this study was a pedestrian survey of the area. Each pedestrian stopped was asked (before and after the lighting was improved there):

If the street lighting was improved, what effect would/did it have on:
1   the number of people walking here at night?
2   the number of bad or unpleasant incidents around here?
3   people's fear of crime around here?
4   the amount of crime committed around here?'

A total of 1,178 respondents were interviewed in three phases: before relighting, then three months afterwards, and 12 months afterwards. Their answers to this question are summarized in Table 8.2, where it can be seen that people seem to have wildly exaggerated expectations of the efficacy of improved street lighting. Had these questions only been asked in the after phase, then some very notable improvements could have been cited (even though they tailed off somewhat over time). By asking for expected improvements – a common approach – expectations dwarf achievements.

Another example comes from Scottish research into the effectiveness of CCTV on crime and the fear of crime (reported in, *inter alia*, Ditton, 1998 and forthcoming; Ditton and Short, 1996, 1998a, 1998b, 1999; Ditton et al., 1999a; Short and Ditton, 1995, 1996, 1998). As part of this overall project, 3,074 pedestrians were interviewed in various parts of Glasgow; 1,206 of them were interviewed in the city centre, where the cameras were to be installed. A third of this city centre group were interviewed nine months before camera installation, another third three months after installation, and the final third 15 months after installation.

At one point, all were asked, 'If CCTV cameras were installed in this street, would you feel more or less safe alone at this time of day?' Rather as in the street lighting example, 61 per cent said that they *would* feel safer (64 per cent, 58 per cent and 61 per cent in the three areas). However, most did not know that the CCTV cameras were there, even after they had been installed. In the second interview, 17 per cent and by the third interview, 20 per cent. But did those who *knew* that there were cameras there feel any safer? No. Actually, 77 per cent of those who *did not* think that they were being watched by cameras said they felt safe when asked 'When you walk along here alone at this time of day, how safe do you feel?' compared to 75 per cent who *did* know they were being watched.

To confound this sort of complexity, those who knew that cameras had been installed were just as likely as those who did not (35 per cent and 36 per cent) to claim that they would feel safer if cameras were installed. In addition, of those who said that they would feel 'safer' if CCTV cameras were installed, 75 per cent had already said that they felt 'safe' walking alone where and when they were interviewed. Finally, those who were aware of the CCTV cameras were no more likely to say that they felt safer, or that they used the streets more often, than those who didn't know about them.

It is not obvious what all this adds up to, but two things seem fairly clear: first, people seem to have exaggerated expectations of the benefits of minor environmental improvements before they are installed (mirroring the benefits promised by those installing them); and second, actual benefits are either perceived to be more muted (the street lighting example) or virtually non-existent (the CCTV example).

Lessons from more recent research

Our research into the methodology of crime and fear-of-crime surveying that we conducted in the late 1990s has persuaded us that much of the data we get from such surveys is a function of the types of questions we ask – rather than of the reality we are probing – and that if we ask different questions we will get quite different 'results'.

One example springs immediately to mind. As part of the preliminary qualitative groundwork for one of our projects, a large number of interviewees were asked for all the different responses that they might have had, at various times, to the experience of or the imagined prospect of criminal victimization. Each listed response was later printed on a separate card, and these cards were given to specially convened focus groups in order for them to rearrange them all into a smaller number of cognate categories. These focus groups all agreed on three categories of response: the 'afraid' category, the 'think' category and the 'angry' category.

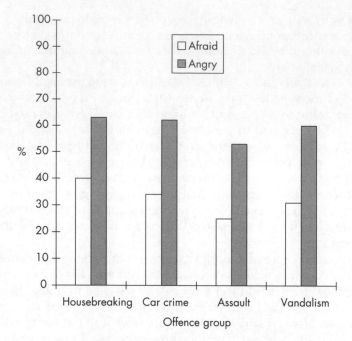

Figure 8.1 *Percentage 'afraid' and 'angry'*

We then operationalized these into closed survey questions, and tested them on a large, pure random sample of 1,629 Scottish adult householders. There was no clearly stated hypothesis at this stage, but if there had been, it would no doubt have been that 'fear' of crime would have been the dominant response. Astoundingly, we discovered, as can be seen from Figure 8.1, that 'anger' was by far the most dominant response.

The overall finding – that 'anger' dominates 'afraid' – was supported in every age and gender category. That is, feelings of 'anger' dominated feelings of 'afraid' across the board. Generally speaking, females were more afraid than males, and they were also angrier. Further, people who were more angry than they were afraid were much more angry than people who were afraid, were afraid. Previous-year victims were admittedly more afraid than were non-victims, but they were also angrier than were non-victims. This had not been discovered before simply because nobody had opened up the research agenda qualitatively and started, as we did, with a blank sheet rather than with somebody else's questionnaire.

We also questioned fear of crime, but not so much to plot its extent and distribution – more to examine the methodological adequacy of the questions normally used. One question we asked was a fairly standard one, slightly rephrased to take account of suggestions from other researchers. It was, 'In your everyday life, are you afraid of someone breaking into your home?' Respondents were given the following response options to choose from. They could answer 'not at all', 'hardly ever', 'don't know', 'some of the time', or 'all the time'.

Table 8.3 *Recoded housebreaking question responses compared*

	not at all + not much	quite a bit + a lot	
Not at all + hardly ever	514 46%	144 13%	658 59%
Some of the time + all of the time	199 18%	256 23%	455 41%
	713 64%	400 36%	1113 100%

Typically, in most crime survey reports, the 'don't know' responses are dropped, and the 'not at alls' are combined with the 'hardly evers', and the 'some of the times' are combined with the 'all the times'. With our data, a sub-set of 1,155 of the 1,629 questioned in total, this summed as 40 per cent (n = 461) who worried about housebreaking, and 60 per cent (n = 694) who did not.

As a check, and unusually for crime surveys, we asked the question again later in the interview, albeit in a slightly different form. This time we asked, 'Could you tell me how worried you are about having your home broken into and something stolen?' Here, the closed response options were 'not at all', 'not much', 'don't know', 'quite a bit' and 'a lot'. The aggregated results were broadly similar, with 40 per cent (n = 461) worrying 'some of the time' or 'all the time' the first time the question was asked, and 35 per cent (n = 402) worrying 'quite a bit' or 'a lot' the second time.

So far, so good. But when the data from the two questions are cross-tabulated (rather than just compared), some unwanted inconsistency of response creeps in. This is illustrated in Table 8.3. Here, it can be seen that 13 per cent of those who worried 'not at all' or 'hardly ever' the first time the question was asked, said that they worried 'quite a bit' or 'a lot' the second time it was asked. Further, 18 per cent who were not worried to start with were worried by the time the second question was asked.

Nearly a third of the sample, then, responded differently when answering pretty much the same question twice within half an hour.

This inconsistency appears greater when the five original closed response options are reinstated, and the original distributions for the two questions are cross-tabulated. This is done in Table 8.4.

It is not altogether clear how this should be dealt with, but rather than complain about 'response inconsistency', more creative use can be made of the data by giving respondents points for worry, and thereby creating worry as a continuum, rather than as a variable with only two values. The points allo-cation system is shown in Table 8.5, where it can be seen that somebody answering 'not at all' to both questions scores 0 points for worry, through to somebody who answers 'all the time' to the first one, and 'a lot' to the second scoring 8.

Figure 8.2 shows the resulting distribution, ranging from the 120 respon-dents (at the bottom) who scored 0 points (didn't worry at all at either

Table 8.4   *Raw housebreaking question responses compared*

	Not at all	Not much	Don't know	Quite a bit	A lot	
Not	120	145	17	44	9	335
at all	10%	12%	1%	4%	1%	29%
Hardly	49	200	15	83	8	355
ever	4%	17%	1%	7%	1%	30%
Don't	9	12	3	1	1	26
know	1%	1%	–	–	–	2%
Some of	68	93	4	165	29	359
the time	6%	8%	–	14%	3%	31%
All the	23	15	–	40	22	100
time	2%	1%	–	3%	2%	9%
	269	465	39	333	69	1175
	23%	40%	3%	28%	6%	100%

Table 8.5   *Raw housebreaking question responses scored*

	Not at all	Not much	Don't know	Quite a bit	A lot
Not at all	0	1	2	3	4
Hardly ever	1	2	3	4	5
Don't know	2	3	4	5	6
Some of the time	3	4	5	6	7
All the time	4	5	6	7	8

question) down to the 22 respondents who worried 'all the time' and 'a lot' (at the top).

We also asked similarly paired questions about both assault and vandalism. They were scored in the same way, and on the three 9-point scales, only between eight and 22 respondents worried 'all the time' and 'a lot' for each. The three scales were then summed to a 24-point scale, where only four respondents (0.3 per cent of the total questioned) worried 'all the time' and 'a lot' about all three possible victimizations. All four were women aged between 43 and 53.

In a sense this poses more questions than answers. First – survey reporting conventions aside – how many respondents can be deemed to be 'worriers' is a matter of choice rather than of fact. Is it just the 22 full-time worriers? Or all but the 120 who never worry? Second, worry clearly is, when the data are analysed in this way a matter of *degree*, and this must logically present not a single policy problem of 'fear', but different degrees of problem and, probably, different problems.

A further example comes from the same research, but this time looking at what are known as the 'indirect' questions on the 'fear' of crime. The believed superiority of indirect measures – also their weakness – is that because crime is not mentioned, any 'automatic' fear response that the word 'crime' triggers will not be included. Typical questions are, 'How safe do you feel walking alone in this area after dark?' and 'How safe do you feel when you are alone in your home at night?' We asked both questions. Notice that the key is how

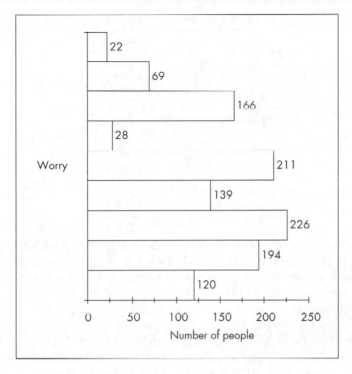

Figure 8.2 *Worry scale for housebreaking fear*

'safe', and the word 'fear' is not mentioned. Because crime is not mentioned either, feelings of unsafety could refer to other matters. 'Alone' and 'after dark'/'at night' are common elements. For both questions respondents who do not go out at night alone are asked how safe they 'would' feel if they did; and those who do not stay in alone at night are asked how safe they 'would' feel if they did.

The offered response range is 'very unsafe', 'unsafe', 'don't know', 'fairly safe' and 'very safe' for both. Typically, the 'very unsafes' are added to the 'unsafes', the 'don't knows' are dropped, and the 'fairly safes' are added to the 'very safe'.

One of the drawbacks of the inevitable need to compress scales into more manageable numbers of responses in this way is the loss of knowledge that can occur. An example comes from the street lighting research referred to earlier, and is of two samples of 130 household-based respondents. The first sample was researched before their street lighting was upgraded, the other afterwards. All are answering the question, 'How safe do you feel walking alone around here after dark?' and the responses are in Table 8.6.

There is a statistically insignificant increase in the number feeling safer, but respondents were actually offered four, not two, possible responses, as shown in Table 8.7.

With this degree of detail, it can be seen that there is a decline in feelings of both safety and unsafety: four fewer people are 'very safe', and three fewer

Table 8.6   Recoded responses

	Before	After
Safe	70 (54%)	71 (55%)
Unsafe	60 (46%)	59 (45%)

Table 8.7   Unrecoded responses

	Before	After
Very safe	35 (27%)	31 (24%)
Fairly safe	35 (27%)	40 (31%)
Bit unsafe	40 (31%)	42 (32%)
Very unsafe	20 (15%)	17 (13%)

people are 'very unsafe'. Now, this was actually a longitudinal study, and the 130 people in each sample were, in fact, the same people. This allows analysis to probe even deeper. By comparing responses case by case, it was found not just that the seven people identified above 'changed', but that 18 did: eight felt safer (one moved from very unsafe to very safe; one from fairly safe to very safe; three from bit unsafe to fairly safe; and three from very unsafe to bit unsafe) but 10 felt less safe (one moving from fairly safe to bit unsafe; one from bit unsafe to very unsafe; two from very safe to bit unsafe; and four from very safe to fairly safe). These subtleties are shown in Figure 8.3.

At one level the simple unanalysed frequency of various responses to such questions has often been used to paint a bleak picture. For example, within the major ESRC-funded Scottish fear-of-crime sample mentioned above, 23 per cent said that they would feel very or fairly unsafe walking alone in their area after dark, and 9 per cent said that they would feel very or fairly unsafe alone at home at night. Respondents were asked a third question, one which is not usually asked. Each was asked, 'How often do you walk around alone locally after dark?' Although only 23 per cent said that they sometimes or often walked around alone locally after dark, the addition of this third question hugely increased the analytic value of the first two. If all three questions are asked, we can then discover whether or not people go out alone at night or stay in, whether or not they feel unsafe when they are out alone, and whether or not they feel unsafe when they in at home alone. The responses can then be combined in one new variable with eight separate values, as shown in Table 8.8.

The possible responses have been scored for apparent degree of 'unsafeness' (with 3 representing greatest 'unsafety'). Notice, first, that the largest number of respondents (row 6: 57 per cent) stay in, feel safe there, and reckon they would feel safe if out. The second-largest number (row 8: 19 per cent) feel safe in, go out, and feel safe when out. A small proportion (row 7: 1 per cent) feel unsafe in, so they go out, and they feel safe there. A larger group (row 2: 14 per cent) would feel unsafe out, but they stay in and feel safe there. So far, this amounts to 91 per cent of the sample who do not really have a problem.

This leaves only 8 per cent of the sample with an 'unsafety' problem. Notice how they do not have a shared one, but constitute four different types

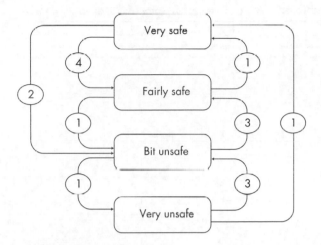

Figure 8.3 *Changes in feelings of 'safety' and 'unsafety'*

Table 8.8 *Relevance of feelings of nightly 'unsafeness'*

Row		Score	n	%
1	Feel unsafe in, stay in, feel unsafe out	3	45	4
2	Feel safe in, stay in, feel unsafe out	1	152	14
3	Feel unsafe in, go out, feel unsafe out	3	14	1
4	Feel safe in, go out, feel unsafe out	2	26	2
5	Feel unsafe in, stay in, feel safe out	2	14	1
6	Feel safe in, stay in, feel safe out	0	627	57
7	Feel unsafe in, go out, feel safe out	1	8	1
8	Feel safe In, go out, feel safe out	0	206	19
	Total		1092	99

Responses do not sum to 1,182 because of missing data.

of problem. Two groups have a minor problem: those who feel safe in, but go out and feel unsafe out (row 4: 2 per cent), and those who would feel safe out, but stay in where they feel unsafe (row 5: 1 per cent). It is a bit flippant to suggest that to reduce 'unsafety', those in the first group should stay in, and those in the second should go out, but why not?

Those in rows 1 and 3 have a more major problem. Those who feel unsafe when they are in, stay in, but would feel unsafe if out (row 1: 45 respondents, 4 per cent of the total), and those who feel unsafe when in, go out, and feel unsafe when out (row 3: 14 respondents, 1 per cent of the total) offer a considerable challenge to those working to enhance community feelings of safety, but notice that they number only 59 respondents out of a sample of 1,092. Further, being a member of either group is not related to gender, age, or past victimization (variables which are traditionally used to explain noticeable feelings of 'unsafety').

It is instructive to see how this group of 59 respondents fared on the earlier generated 24-point worry scale. Table 8.9 has the data.

Table 8.9   *Direct victimization 'worry' and indirect 'unsafeness'*

Worry scale	Row 1 'unsafe'	Row 3 'unsafe'	Rows 1+3 'unsafe'
1	1	1	2
2			
3			
4		2	2
5	5		5
6	6		6
7	3		3
8	1	1	2
9			
10	3		3
11	2	1	3
12	4	2	6
13	3		3
14	2	1	3
15	2		2
16	4	3	7
17			
18	5		5
19	4	2	6
20			
21	1		1
22			
23	1	1	2
24	1		1
Total	45	14	59

Incidentally, of the four respondents who, earlier analysis indicated, on a 24-point scale worried 'all the time' and 'a lot' about becoming a victim of assault, housebreaking and vandalism (all four were women aged between 43 and 53), one was a woman who feels unsafe when she is in, stays in, and would feel unsafe if out (i.e. in row 1 for 'unsafety': at the bottom of the second column of Table 8.9), but the other three were all women who have no problem on the 'unsafeness' scores: all would feel unsafe out, but they stay in and feel safe there, that is, they are in row 2 of Table 8.8.

Putting the two scales together, and viewing the situation rather unsympathetically, we have only one person who is 'really' worried. But is this a social problem or a medical one?

Conclusion

Our own experience – garnered in over a decade of research into the issue – seems to point to a number of possibly general conclusions. First, respondents are prepared to say they 'fear' crime in proportions that are remarkably consistent over time and place, and stubbornly resistant to amelioration by environmental and other improvements. This suggests, to us, that the so-called fear of crime is – to an unknown degree – a function of the types of questions that are asked, and the way that they are posed. Second, anger about crime rather than fear of crime is the prime reaction to both the prospect of and reality of criminal victimization experienced by most people. Naturally enough, surveys that fail to ask how angry people feel about crime victimization will gather no data on it. Third, and perhaps most general of all, the fear of crime is researched chiefly by closed, positivistically informed survey instruments. These questionnaires rarely exhibit novelty, and depend chiefly on incestuous and uncritical filching of questions from the British Crime Survey. Following the period of introspective contemplation for fear of crime research that we recommended at the beginning of this chapter, a period of open, qualitatively based re-examination of the general population's attitudes to crime is long overdue.

Suggested readings

Glaser, B. and Strauss, A. (1967) *The Discovery of Grounded Theory*. New York: Free Press.
Schuman, H. and Presser, S. (1996) *Questions and Answers in Attitude Surveys*. London: Sage.

References

Ditton, J. (1998) 'Public support for town centre CCTV schemes: myth or reality?', in C. Norris, J. Morgan and G. Armstrong (eds) *Surveillance, Closed Circuit Television and Social Control*. Aldershot: Ashgate. Chapter 12.

Ditton, J. (forthcoming) 'Crime and the city: public attitudes towards open-street CCTV in Glasgow', *British Journal of Criminology*.

Ditton, J. and Farrall, S. (forthcoming) 'Fear of burglary: refining national survey questions for use at the local level', *International Journal of Police Science and Management*.

Ditton, J. and McNair, D. (1994) 'Public enlightenment', *The Surveyor*, 181 (5282): 18–19.

Ditton, J. and Nair, G. (1994) 'Throwing light on crime: a case study of the relationship between street lighting and crime prevention', *Security Journal*, 5 (3): 125–132.

Ditton, J. and Short, E. (1996) 'Yes – CCTV affects crime!', *CCTV Today*, 3 (4): 14–18.

Ditton, J. and Short, E. (1998a) 'When open street CCTV appears to reduce crime – does it just get displaced elsewhere?', *CCTV Today*, 5 (2): 13–16.

Ditton, J. and Short, E. (1998b) 'Evaluating Scotland's first town centre CCTV scheme: Airdrie's crime data and Airdrie's offenders', in C. Norris, J. Morgan and G. Armstrong (eds) *Surveillance, Closed Circuit Television and Social Control*. Aldershot: Ashgate. Chapter 8.

Ditton, J. and Short, E. (1999) 'Yes, it works – no, it doesn't: comparing the effects of open-street CCTV in two adjacent town centres', *Crime Prevention Studies*.

Ditton, J., Nair, G. and Phillips, S. (1993) 'Crime in the dark: a case study of the relationship between streetlighting and crime', in H. Jones (ed.) *Crime and the Urban Environment: The Scottish Experience*. Aldershot: Avebury. pp. 99–114.

Ditton, J., Nair, G. and Bannister, J. (1996) 'The cost-effectiveness of improved street lighting as a crime prevention measure', *Lighting Journal*, 61 (4): 251–256.

Ditton, J., Farrall, S., Bannister, J. and Gilchrist, E. (1998) 'Measuring fear of crime', *Criminal Justice Matters*, 31 (Spring): 10–12.

Ditton, J., Bannister, J., Gilchrist, E. and Farrall, S. (1999a) 'Afraid or angry? Recalibrating the fear of crime', *International Review of Victimology*, 6 (2): 83–99.

Ditton, J., Short, E., Norris, C. and Armstrong, G. (1999b) *The Effect of Closed Circuit Television Cameras on Recorded Crime Rates, and the Fear of Crime in Glasgow*. Edinburgh: Scottish Office Central Research Unit.

Ditton, J., Farrall, S., Bannister, J., Gilchrist, E. and Pease, K. (1999c) 'Reactions to victimisation: why has anger been ignored?', *Crime Prevention and Community Safety: An International Journal*, 1 (3): 37–54.

Ekblom, P., Law, H. and Sutton, M. (1996) *Safer Cities and Domestic Burglary*. Home Office Research Study No. 164. London: Home Office Research and Statistics Directorate.

Farrall, S., Bannister, J., Ditton, J. and Gilchrist, E. (1997a) 'Measuring crime and the "fear of crime": findings from a methodological study', *British Journal of Criminology*, 37 (4): 657–678.

Farrall, S., Bannister, J., Ditton, J. and Gilchrist, E. (1997b) 'Open and closed questions', *Social Research Update*, 17.

Farrall, S., Ditton, J. and Bannister, J. (submitted) 'Quantitative testing of the fear of crime: the apparent demise of age and gender as key explanatory variables'.

Farrall, S., Bannister, J., Ditton, J. and Gilchrist, E. (forthcoming) 'Social psychology and the fear of crime: re-examining a speculative model', *British Journal of Criminology*.

Gilchrist, E., Farrall, S., Bannister, J. and Ditton, J. (1998) 'Women and men talking about the "fear of crime": challenging the accepted stereotype', *British Journal of Criminology*, 38 (2): 283–298.

Horne, C. (1996) 'The case for: CCTV should be introduced', *International Journal of Risk, Security and Crime Prevention*, 1 (4): 317–326.

McNair, D. and Ditton, J. (1995) 'Does anybody know the cost of street lighting?', *Lighting Journal*, 60 (5): 299–300.

Nair, G. and Ditton, J. (1994) '"In the dark, a taper is better than nothing": a one year follow-up of a successful streetlighting and crime prevention experiment', *Lighting Journal*, 59 (1): 25–27.

Nair, G., Ditton, J. and Phillips, S. (1993) 'Environmental improvements and the fear of crime: the sad case of the "pond" area in Glasgow', *British Journal of Criminology*, 33 (4): 555–561.

Nair, G., McNair, D. and Ditton, J. (1997) 'Street lighting: unexpected benefits to young pedestrians from improvement', *Lighting Research & Technology*, 29 (3): 143–148.

Short, E. and Ditton, J. (1995) 'Does CCTV affect crime?', *CCTV Today*, 2 (2): 10–12.

Short, E. and Ditton, J. (1996) *Does Closed Circuit Television Prevent Crime? An Evaluation of the Use of CCTV Surveillance Cameras in Airdrie Town Centre*. Edinburgh: Scottish Office Central Research Unit.

Short, E. and Ditton, J. (1998) 'Seen and now heard: talking to the targets of Open Street CCTV', *British Journal of Criminology*, 38 (3): 404–428.

9

SNEAKY MEASUREMENT OF CRIME AND DISORDER

Jeanette Garwood, Michelle Rogerson and Ken Pease

Contents

Various policy developments of recent years inform current strategies for the control of crime and disorder. The philosophy of problem-oriented policing (Goldstein, 1990, 1993) emphasizes the problem-solving function of police and other agencies at the expense of purely technical measures of police work such as response time. In the UK, the highly influential Morgan Report (1991) introduced the term 'community safety', which has since gained wide currency. Morgan advocated that local authorities be assigned statutory responsibility for crime prevention/community safety. Coming from a wholly different perspective from Goldstein, the fundamental implications of the approach were the same, namely that addressing factors which made citizens be or feel unsafe required a broader understanding of the local determinants of problems than was offered by a traditional approach in which criminal justice was central. While the Morgan Report was shelved by a Conservative administration intent upon de-emphasizing local government, its approach nonetheless informed the major crime reduction initiatives of the early 1990s. Commentaries on these developments, their problems and their tendency to become politicized now exist in the criminology literature (see Crawford, 1996; Gilling, 1997; Hughes, 1998; Marlow and Pitts, 1996).

With the return of the Labour administration in 1997, the ideas under-pinning Morgan were revived in modified form and legislated as the Crime and Disorder Act 1998. The essential difference between the provisions of the Act and what Morgan had envisaged was that police and local authorities have equal and joint responsibility for assessing and achieving community safety. The exclusive local authority responsibility for crime prevention proposed by Morgan has been replaced by local 'responsible authorities', including both local authorities and the police acting in concert. At a meeting of the Thames Valley Partnership on 11 March 1998 the then Minister of State at the Home Office, Alun Michael, expressed the relationship as 100 per cent police responsibility and 100 per cent local authority responsibility. Section 6 of the Act requires local 'responsible authorities' to:

(1)   . . . formulate and implement, for each relevant period, a strategy for the reduction of crime and disorder in the area.
(2)   Before formulating a strategy, the responsible authorities shall –

carry out a review of the levels and patterns of crime and disorder in the area;
prepare an analysis of the results of that review;
publish in the area a report of that analysis; and
obtain the views on that report of persons or bodies in the area . . .

A strategy shall include –

objectives to be pursued by the responsible authorities, by co-operating persons or bodies . . .
long-term and short-term performance targets for measuring the extent to which such objectives are achieved.

While the rhetoric of partnership and problem-solving is universal, the reality of strategies for change and their evaluation give pause for thought. Experience of the Safer Cities programme, Single Regeneration Budget projects and related programmes suggests that such review and performance will have considerable shortcomings (see Pawson and Tilley, 1997). This will be especially true for the first three-year cycle of the audit–strategy–review sequence envisaged by the Crime and Disorder Act, wherein the process is new to all concerned. Indeed the writers' understanding of the first audit was of a frantically rushed and immediately politicized process familiar from the earlier schemes covered by the writers cited above.

One can thus summarize and infer from previous work and early experience of the 1998 Act that there are several reasons for the typical past short-comings which are likely to be reproduced as shortcomings of any future review and evaluation undertaken by local responsible authorities in relation to community safety. The major problems are believed to be as follows: First, the absence of appropriate measurement carries with it local political advantages. It means that resources can be directed to politically popular causes or to a councillor's own ward, without the need to argue against data that demonstrate where a problem really exists. Second, there is a conceptual gap between proposed action and envisioned outcomes. In Pawson and Tilley's terms, there is a lack of articulation of contexts, mechanisms and outcomes (see Tilley in Chapter 5 for a summary of these). (It is interesting that this now seems to be acknowledged, for example in the Home Office guidance notes for the fourth round of its CCTV Challenge competition.) Third, currently available measures are truly inadequate and can only be remedied at great cost. A detailed study of the Merseyside Partnership (funded by the Single Regeneration Budget) shows that a major factor limiting its success was the lack of appropriate measurement, together with the lack of articulation between that which was measured and the mechanism by which community safety was to be delivered (Buck, 1998).

The problem with existing local measures of crime and disorder

How do local responsible authorities go about reviewing patterns of crime and disorder? Very early experience of the Act in operation suggests that some make only token gestures, allowing full rein to local lobbying and political partiality in determining the crime audit. Those who seriously engage with the obligation undertake some combination of analysis of reported crime and calls for service, as recorded by the police; remembered crime, derived from a victimization survey; and some anecdotal accounts (perhaps formalized in focus groups). The shortcomings of these measures are well known (e.g. see Maguire, 1997).

While unreported crime may not massively distort the picture provided by recorded crime data (although for crimes such as domestic violence and hate crime they might well: see Farrell and Buckley, 1999), there are other limitations to these data. They are not usually helpful in clarifying time, place and circumstance in enough detail to generate focused action. The geocoding of recorded crime data (see for example Hirschfield and Bowers, 1997) is in its infancy in many police forces, with limited back record conversion even in those areas where it is in place. The identification of crime hot spots is correspondingly imperfect. Manual analysis of police crime reports will reveal that often the location is recorded so imperfectly that it is impossible to geocode at all. This is especially true of vehicle crime.

Victimization surveys are expensive, estimated as a minimum of £30,000 per 1,000 respondents adequately sampled and subjected to a 45-minute interview. The use of national victimization data using the British Crime Survey is possible, but does not provide the necessary local resolution of crime and disorder without compromising victim anonymity. The only police force to supplement British Crime Survey data with a local sample spent some £40,000 on doing so, and the results yielded have so far not been of such a kind as to inform local action. The presumption must be that 'long-term and short-term performance targets' should be explicitly linked to the data from the crime and disorder review. If so, this makes the victimization survey impossibly expensive, given that it needs to be repeated to show whether performance targets have been achieved. The British Crime Survey, whilst repeated every two years, cannot overcome this problem as the time that elapses between data collection and access to the survey makes it difficult to link trends in crime to local initiatives. Thus, in short, the basic limitations of crime victimization surveys are cost and lack of precise local relevance.

The role of anecdote in characterizing area reputation is considerable. Anecdote is just that, although disproportionately important in the politicized atmosphere which often surrounds community safety. Anecdote may serve to indicate that there is a problem, but fails to demonstrate any generalizable principles on level or extent of crime and disorder.

In short, the Crime and Disorder Act places the local 'responsible authorities' in a difficult position. None of the measures of crime and disorder to which they will have regular recourse provides adequate means to conduct an accurate yet tolerably cheap review of crime and disorder.

Do crime and disorder go together?

There is a need to be clear about the reasons why crime and disorder have become linked in public policy. Why, for example, was the Crime and Disorder Act 1998 not the Criminal Justice Act 1998? Something happened to policy assumptions in the 1990s which yoked crime and disorder together.

Skogan (1990: 10) notes that 'neighbourhood levels of disorder are closely related to crime rates, to fear of crime, and the belief that neighbourhood crime is a serious problem'. The presumption (consistent with Skogan's and other evidence) is that the relationship is a causal one, with disorder driving out community controls and the people who might exercise them, leading to crime escalation. The confluence of crime and disorder as greater or lesser symptoms of the same underlying problem means that measurement issues need to be revisited. It means, *inter alia*, that traces of disorder do not need to be individually linked to measures of crime. Measures of disorder do not need to be markers of something else. Indeed, a measurement strategy which limited itself to crime indices or rejected disorder measures that were precursors rather than correlates of crime would necessarily be inadequate to the purposes of the Act.

A supplementary measurement issue is how people behave in disorderly areas. Where the elderly go and when, for example, may be taken as proxies for areas which are safe or otherwise. In the world of community safety practitioners, an increase in the times at which older people visit town centres may be an important outcome measure.

Trace measures of crime, disorder and use

While we propose a set of measures of crime and disorder, we must assert that we are not so naïve as to suppose that the measures to be outlined provide, alone, some sort of royal road to the measurement of crime and disorder. The methods proposed have certain advantages in cost, short lead time to yield measures and, in our view crucially, the focusing of the minds of implementers upon the immediate means by which they would conclude that their approach was working through the mechanism whereby it was intended to work. Triangulation (the use of several different methods of data collection and analysis) is always a good idea, and where methods are cheap and accessible, it is likely to be especially fruitful. One aspect of the present situation which makes the development of these methods a matter of urgency is the novelty of the process through which community safety practitioners are required to go. With the best will in the world, the first crime audit will be flawed, and as a result practitioners will be faced at some time during the first three year cycle (1998–2001) with having to address problems highlighted in the audit which in reality turn out to be not so serious; or be faced with more pressing problems thereby forgoing the evaluative underpinning of their work. It is all about addressing priorities in the context of limited resources and time. The availability of cheap and simple indices would mean that change to the more pressing problems could be made without completely sacrificing the evaluative element.

Trace measures: a brief history

The use of unobtrusive or non-reactive measures was initially documented and advocated by Webb et al. in 1966, and further developed by Webb and others in 1981. An unobtrusive measure is one which is collected by a method which in no way affects the phenomenon under study. This can include participant observation, which is the observation of individuals without them knowing the true reason for the presence of the researcher. In this chapter the focus is on the use of one form of unobtrusive measure – the trace measure. Traces are 'things' produced by individuals or groups of individuals which a researcher may use as an indicator of some form of social behaviour. For example, market researchers find it very useful to know which stations potential customers tune their car radios to. Street surveys are a possibility, asking car drivers which station they usually listen to; there are problems with the social desirability of such an approach, where the over-35s might insist that they listen to Classic FM, whilst the under-25s claim that they listen to local popular music stations. The trace measurement alternative would be to pay garage companies, servicing and repairing cars, to have their staff check to which station each car radio was tuned on arrival (Anonymous, 1962). Other researchers have used the content of waste bins to monitor the consumption of take-away foods by young women in catered university accommodation (cited in Rathje, 1984), and to measure the wear and tear of floor tiles as an indicator of the most popular exhibits in museums (Duncan, 1963, cited in Webb et al., 1966). More recently it has been used to explore the demarcation of neighbourhoods and territories in Northern Ireland (Robson, 1993). Currently, trace methods – such as recording the placing and content of litter bins; noting noise levels and losses and damage to books and journals – are being undertaken within a university library. The goal is to design appropriate interventions to reduce losses, and to test the hypothesis that the presence of low-level delinquency is a key to losses and damage in particular subject areas and physical locations (Garwood, in progress). The aim in all of these cases has been to study behaviours which would be very likely to have response bias under more direct forms of measurement or where direct questioning is used, as is the case with crime surveys.

Unobtrusive measurement has several crucial advantages over the alternatives. It is normally cheap (often utilizing data already collected by authorities), it is not liable to response bias, and it is seldom attached to any single individual, reducing the ethical problems associated with invasion of privacy. For example, injectable drug use can be monitored by counting or weighing the contents of street bin collection points, and monitoring change across time, and comparing this with reports of needles and other injectables paraphernalia found by street cleaners, or with public complaints. It will not always be conclusive when taken alone, but will be suggestive. Triangulation, involving police recorded crime data and unobtrusive measurement, may represent the best bargain for the review and performance measurement process required by the Crime and Disorder Act. A further advantage of these methods is that the process of selecting measures is an aid to inter-agency bonding, helping responsible bodies to work together to tackle crime and disorder. The need, at this point, is to devise and refine modern unobtrusive measures for this purpose.

Modern traces of crime and disorder

What might modern measures of crime and disorder be? Some examples are given below:

- Alcohol consumption in public: traces of this can be found in litter bins in parks and public gardens, with the kind of drink (Alcopops, cheap wine, sherry, cider or lager) giving a clue as to who may be doing the drinking.
- Seat repair costs in football grounds: it is likely that such damage will be higher in the away section of grounds, since travelling supporters are typically more committed and volatile. It is predicted that damage will be highest in the seats at the territorial divide between home and away fans, since fans who choose such seats are those most interested in taunts and confrontation.
- Other indicators of vandalism might include damage, or the rate of repairs to buses (see Sturman, 1976), bus shelters and telephone boxes.
- Non-standard maintenance costs in schools and between tenancies in council-owned housing will give an indication either of damage inflicted by tenants or to properties that are empty between tenancies.
- The paraphernalia of smoking heroin is distinctive, with matches and burned foil. The utility rooms of multi-storey blocks and the toilets of pubs in which drug use is extensive have much debris of this kind. Perhaps the smoking of cannabis can be indexed by calculating the ratio of rolling tobacco to cigarette papers sold: it is likely that far more papers will be

bought than could be accounted for by the amount of rolling tobacco. Clubs which host much amphetamine use may be characterized by a high rate of sale of water and soft drinks.

- Indicators of incivility include visible abuse of the orange badge scheme. This can be seen from different viewpoints. First, the number of non-orange-badge holders parking in disabled bays can be counted (some authorities already book cars for this); second, illegal use of orange badges by trade vehicles (such as those of builders) act as a further indicator of incivilities. In a sense, this reflects the thinking of the massively influential 'broken windows' hypothesis (Kelling and Coles, 1997), whereby the existence of a broken window which remains unreplaced is a marker for indifference to the locale.

- Positive markers for improvement of feelings of safety in an area are also possible. Increased sales of bus passes, and take-up of free or reduced-price bus passes amongst the elderly could, along with records of the concessionary fares, be very useful for this purpose. A further indicator is increased takings from streetside pay and display units; meter takings were used in the US to measure the success of a newspaper strike as long ago as 1963 (Mindack et al., 1963).

One step further, at the time of writing the Chief Constable of Thames Valley sees it as extremely important to determine the level of 'social glue' (Pollard, personal communication). There are many ways of checking this – for example levels of redirection of deliberately misdirected mail, levels of pension collection on behalf of a pensioner, levels of use of schools for communal purposes, extent to which people return the wheely-bins of their neighbours after refuse collection has taken place. Changes in levels of report of broken or failed street lamps, and uneven paving, or complaints to Highways that pavements have not been repaired properly or are dangerous, may give a clue to the levels of feelings of perceived control and social cohesiveness. Such indicators of 'social glue' may be particularly useful in establishing the presumptive mechanism of change. Some scattered work of a similar nature has already been undertaken, as is indicated by the occasional references given above. Other examples include Coleman's (1985) measures of dog faeces and the like as an indication of area breakdown (although the way in which she combined and interpreted these is open to question). Sturman (1976) links levels of damage in Manchester buses to route, staffing and bus design, in order to make prevention recommendations. Rates of infective hepatitis have been used as proxies for the amount of drug use by injection (Joint Committee on the New York Drug Law Evaluation, 1978).

Some of these traces of crime, disorder and incivilities will already be on record, but may not have been collated (e.g. parking in disabled bays without an orange badge); others will need collecting and codifying, such as infective hepatitis figures; others, such as debris from heroin smoking, will need to follow on from the development of methodologies for data collection, such as those of the North American 'Project du Garbage' (Hughes, 1984; Rathje and Hughes, 1984).

One example may be given of actual use of such a technique, albeit very modest in scale. An undergraduate student at the University of Huddersfield, Colette Felvus, under the supervision of two of the authors, sought to count heaps of toughened glass in the car park of the Meadowhall Shopping Centre, Sheffield. Such heaps, found by the roadside or in car parks, are almost certainly indicators of theft from cars. She found the vast majority of such heaps in one small section of the car park, characterized by the absence of passing people on foot, and easy access to waste land. The number of such piles of glass, when annualized, exceeded the total number of reported thefts from cars in the entire police division in which the car park is sited. It is well known from all sweeps of the British Crime Survey that theft from vehicles is greatly under-reported, but to find that one technique (window breaking) in one car park exceeds the recorded total for a division gives dramatic evidence of the extent of under-report.

Assessing validity of trace measures

Of course, where possible, trends in unobtrusive measures must be checked against other readily available data. For example, when there is an increase in replacement of windows or doors in a school, it must be ascertained that such turnover of these structural items is due to attempted break-ins, or to actual break-ins and vandalism, not a replacement programme in that organization. Some expected proportion of normal to abnormal replacement is likely to be used by actuaries, but this needs to be calculated in order to obtain an unobtrusive indicator.

A further way of checking the meaning of unobtrusive measures would be to exploit the fledgling results of researchers such as Hirschfield and Bowers (1997). By going to the areas highlighted in their data as crime hot spots, it would be possible to track the proposed unobtrusive measures, and to establish whether they do indeed discriminate between the crime hot spots and much lower-rate crime areas. Having established that such measures do discriminate, it is possible to transplant the method to areas which have not been profiled. This is not the only method of checking, merely one to serve as illustration. Researchers have, without the sophisticated software of contemporary research, mapped indirect indicators in the past. For example, Shaver et al. (1975) found that fire false alarms could be predicted by environmental risk factors, including the vicinity of empty property, large, neglected warehousing, and less well travelled streets.

As already mentioned, anecdote is important to the local view of criminal and disorderly behaviour. Supported by quantitative data the value of local knowledge and belief is considerable. For example, certain parks and recreational grounds have a reputation among parents as unsafe places for children to play. Bad reputations result from suspected drug use (especially injectables), or solvent abuse, or under-age drinking. As a result these parks are underused. By the investigation of the content of park litter bins in strategic places local knowledge and belief can be linked to trace measures. Clearing the litter bins from these areas, and looking over their content in a systematic manner, can provide much information on whether there is physical evidence to support the anecdote (Hughes, 1984). Since the majority of park keepers are currently equipped with sturdy protective gloves, some hazards are already recognized. Bins must be emptied; the addition here is that material is bagged in such a way that the content of a particular bin can be identified. So, if a park has a reputation for drug use, then needles and syringes might be expected (heroin smoking is likely to need some shelter); for under-age drinking, cider bottles and cans might be in high volume; for street drinkers, the bottles are perhaps more likely to be fortified wines and cheap sherries. The use of trace measures confirms or contradicts local reputations, and in either event public space is liberated for use by children and adults. Such data collection may even help local inhabitants to feel that their worries and complaints are beginning to be taken seriously, and contribute to local cohesiveness. There is of course the danger that offenders will move rather than stop uncivil or illegal behaviour. Tracking litter bin content beyond the original area of interest might be a method for tackling the question of displacement.

Conclusion

While all the arguments advanced for trace measurement above are genuine, it would be disingenuous to omit one which is central to our thinking. That is that research should, in the colourful American phrase, get down and dirty. The use of abstract nouns like community cohesion, disorder and the like allows such slippage in interpretation that partnerships can believe they are talking about the same things but in fact be talking past each other. Nothing is more specific than an Alcopop bottle in a town square litter bin, shards of glass outside a particular club, and a Ford Escort without an orange badge parked in a disabled bay. If these are the signs of disorder, incivility and the like, they are the lowest and most concrete common denominator of what community safety organizers must change. They provide the vocabulary and unit of measurement closest to the realities of the street, store and club. They do not allow retreat to the abstract. They require full focus on how you will know you have changed an area. In that sense, they are as much about the resocialization of police and other community safety practitioners away from the rhetoric of arrests and rehabilitation to the specifics of change of the world that people must inhabit.

Suggested readings

Pawson, R. and Tilley, N. (1997) *Realistic Evaluation*. London: Sage.
Webb, E.J., Campbell, D.T., Schwartz, R.D. and Sechrest, L. (1966) *Unobtrusive Measures: Nonreactive Research in the Social Sciences*. Chicago: Rand McNally.
Webb, E.J., Campbell, D.T., Schwartz, R.D., Sechrest, L. and Grove, J.B. (1981) *Nonreactive Measures in the Social Sciences*. Boston: Houghton Mifflin.

References

Anonymous (1962) 'Z-Frank stresses radio to build big Chevy dealership', *Advertising Age*, 33: 83.

Buck, W. (1998) 'Partnership in community safety: The Merseyside partnership'. PhD thesis, University of Manchester.

Coleman, A. (1985) *Utopia on Trial*. London: Hilary Shipman.

Crawford, A. (1996) *The Governance of Crime*. Oxford: Clarendon Press.

Farrell, G. and Buckley, A. (1999) 'Evaluation of a police domestic violence unit: repeat victimization as a performance indicator', *Howard Journal of Criminal Justice and Crime Prevention*, 38 (1): 41–53.

Garwood, J. (in progress) 'Trace measures and low level delinquency in a university library setting'. University of Huddersfield.

Gilling, D. (1997) *Crime Prevention*. London: UCL Press.

Goldstein, H. (1990) *Problem-Oriented Policing*. Philadelphia, PA: Temple University Press.

Goldstein, H. (1993) 'Confronting the complexity of the police function', in L. Ohlin and F.J. Remington (eds) *The Tension between Individualization and Uniformity*. Albany: State University of New York Press.

Hirschfield, A. and Bowers, K.J. (1997) 'The development of social, demographic and land use profiler for areas of high crime', *British Journal of Criminology*, 37: 103–120.

Hughes, G. (1998) *Understanding Crime Prevention*. Buckingham: Open University Press.

Hughes, W.W. (1984) 'The method to our madness: the garbage project methodology. Special issue on household refuse analysis, theory, method and applications in social science', *American Social Scientist*, 28: 41–50.

Joint Committee on the New York Drug Law Evaluation (1978) *The Nation's Toughest Drug Law: Evaluating the New York Experience: Final Report*. Washington, DC: National Institute of Law Enforcement and Criminal Justice.

Kelling, G. and Coles, C. (1997) *Fixing Broken Windows*. New York: Free Press.

Maguire, M. (1997) 'Crime statistics, patterns, and trends: changing perceptions and their implications', in M. Maguire, R. Morgan and R. Reiner (eds) *The Oxford Handbook of Criminology*, 2nd edition. Oxford: Clarendon Press. pp. 135–188.

Marlow, A. and Pitts, J. (eds) (1996) *Planning Safer Communities*. Lyme Regis: Russell House.

Mindack, W.A., Neibergs, A. and Anderson, A. (1963) 'Economic effects of the Minneapolis newspaper strike', *Journalism Quarterly*, 40: 213–218.

Morgan Report (1991) *Safer Communities: the Local Delivery of Crime Prevention through the Partnership Approach*. London: Home Office.

Pawson, R. and Tilley, N. (1997) *Realistic Evaluation*. London: Sage.

Rathje, W.L. (1984) 'The garbage decade. Special issue on household refuse analysis, theory, method and applications in social science', *American Social Scientist*, 28: 9–30.

Rathje, W.L. and Hughes, W.W. (1984) 'Introductory comments, special issue on household refuse analysis, theory, method and applications in social science', *American Social Scientist*, 28: 5–8.

Robson, C. (1993) *Real World Research*. Oxford: Blackwell.

Shaver, P., Schurtman, R. and Blank, T.O. (1975) 'Conflict between firemen and ghetto dwellers: environmental and attitudinal factors', *Journal of Applied Social Psychology*, 5: 240–261.

Skogan, W.G. (1990) *Disorder and Decline*. New York: Free Press.

Sturman, A. (1976) 'Damage on buses: the effects of supervision', in P. Mayhew et al. (eds) *Crime As Opportunity*, Home Office Research Study No. 34. London: HMSO.

Webb, E.J., Campbell, D.T., Schwartz, R.D. and Sechrest, L. (1966) *Unobtrusive Measures: Nonreactive Research in the Social Sciences*. Chicago: Rand McNally.

Webb, E.J., Campbell, D.T., Schwartz, R.D., Sechrest, L. and Grove, J.B. (1981) *Nonreactive Measures in the Social Sciences*. Boston: Houghton Mifflin.

PART III

EXPERIENCING CRIMINOLOGICAL RESEARCH

The importance of reflexivity

Reflection on the decisions which have been taken in research and on the problems which have been encountered is an essential element of doing research. In fact it is often the case that a reflexive account is published as part of a research report or a book. Typically, such an account covers all phases and aspects of the research process. For example, it will outline and discuss how a research problem came to take the shape that it did, how and why certain cases were selected for study and not others, the difficulties faced in data collection and the various influences on the formulation of conclusions and their publication. Reflexive accounts should not be solely descriptive but should also be analytical and evaluative. Reflexivity is not a self-indulgent exercise akin to showing photographs to others to illustrate the 'highs' and 'lows' of a recent holiday. Rather, it is a vital part of demonstrating the factors which have contributed to the social production of knowledge.

Research findings and conclusions are not 'things' that are lying around waiting to be picked up by an investigator; they are the outcome of research decisions which are taken at different stages and of the factors that influence these, including factors external to and out of the control of the investigator. Research design is an exercise in compromise whereby the investigator seeks to trade off the strengths and weaknesses of different methods. But it is not possible to escape the reality that even the best laid plans and designs have to be actualized in social, institutional and political contexts which can have a profound effect on the outcome of research. Giving recognition to this is important on two counts: first, it allows some assessment to be made of the likely validity of conclusions; and it encourages us to reflect critically on what comes to pass as 'knowledge', how and why. This latter aspect is one hallmark of critical social research. The contribution of reflexivity to assessment of validity and also to critical social research will be discussed later.

Research as a social activity

Three assumptions underpin the concern of this part with experiencing criminological research. The first is that research is a social activity. Criminology is not like those physical sciences in which researchers study and engage inanimate objects. In the main, social researchers are concerned with individuals – although not always at first hand – and these are people with feelings, opinions, motives, likes and dislikes. What is more, typically, social research is a form of interaction. Criminologists should easily recognize this because one influential theoretical approach within the discipline – interactionism – emphasizes that what comes to be recognized as 'criminal' can be the outcome of interactions in the processes of the criminal justice system. Therefore it should come as no surprise that what comes to pass as 'knowledge' can be the outcome of interactions in the research process.

Research and politics

Criminological research is not just a social activity, it is also a political activity. It involves some form of relationship between the subjects of research and the investigators, but there are also others who have an interest. The range of stakeholders typically includes sponsors of research, gatekeepers who control access to sources of data and the various audiences of research findings. These audiences include the media, policy-makers, professionals working in the criminal justice system, politicians and academics. Gatekeepers may have a formal role and legal powers to restrict access (for example, a prison governor) or they may be able to deny access by informal means (for example, by continually cancelling appointments). Sponsors of research include government departments, especially the Home Office, institutions of criminal justice, such as the police or the legal profession, and pressure groups such as the Howard League or NACRO. Each of these stakeholders has interests to promote and interests to protect. Also, each has differential levels of power with which to promote and protect such interests. The exercise of such power is ingrained in the research process from the formulation of problems through to the publication of results.

Research and politics connect in differing ways. For example, politics can have an impact on the course which research takes and also on its outcome. The kind of research which is funded and the ways in which research problems and questions are framed are very much influenced by sponsors. Often they are interested in policy relevance (in their terms) and insist on a formal customer–contractor relationship in which 'deliverables' are clearly specified. How research activity takes place is also dependent on the willingness of subjects to take part and on whether gatekeepers give access to subjects – or other data sources – in the first place (see Chapter 12 for a discussion of the factors which can affect the progress of prison research).

A second way in which politics and research connect is in the differing ways in which the activity of research and its outputs contribute to politics. One important way in which this occurs in criminology is in the conduct of policy-related research. Such research can take a variety of forms but one which has contributed substantially to the formulation and implementation of criminal justice policy is evaluation research (see Chapters 5, 6 and 7). Sometimes this kind of work is known as administrative criminology because of its contribution to the administration and management of the criminal justice system. However, criminologists who represent a critical approach see such work not solely as contributing to policy, but, more importantly, as justifying policy. In this sense they look upon policy-related research as playing a political role in mechanisms of social control and not as benign, value-free contributions to administration and management. In Chapter 10 Barbara Hudson outlines some of the major strands within critical social science. These include the analysis of ideologies which underpin social structures (such as law and order ideologies) and the challenging of these ideologies with the aim of replacing them. In this way, criminology becomes a political activity.

Research and ethics

Research is not just a social and political activity but also an ethical activity. In fact, it is *because* it is a social and political activity that it is an ethical activity. Ethics is about the standards to be adopted towards others in carrying out research. Sometimes these standards are mandatory to the practice of research, for example in the conduct of certain kinds of medical research, whereas in other contexts and disciplines they are merely guidelines. Sometimes they are formally expressed in professional codes of conduct such as in the ethical code of the British Psychological Society, whereas in other disciplines there is a much less formal body of custom and practice.

One ethical principle which is often expressed in social research is that of informed consent. This can be rather elastic but basically it refers to the principle that the subjects of research should be informed of their participation in research, which may be taken to include giving information about possible consequences of participation. Further, it includes the belief that subjects should give their consent to participation, and its possible consequences, prior to their inclusion. Another principle which is sometimes propounded is that no person should be harmed by research: for example, that the introduction of 'experimental treatment' in some styles of research should not cause physical or psychological damage to subjects or, perhaps, disadvantage some individuals in comparison with others.

Matters of ethics interact with the pursuit of validity and also with the political dimensions of research. If the principle of informed consent is applied in full and in such a way that subjects are aware of all aspects of research, including its purpose, it is highly likely that they will behave or react in ways in which they would not normally do. Such reactivity on the part of subjects is a threat to the validity of findings. Further, the challenging of the ideological positions of certain groups in society – perhaps with a view to replacing them with others – is a central aim of some forms of research, especially critical research. However, this inevitably involves doing harm to the interests of such groups. In this way, the fundamental aims of critical research can come face to face with the ethical principle that research should not harm or damage individuals or groups of individuals.

The case for reflexivity

It has been emphasized that reflexivity on the part of a researcher is a vital part of criminological research. This is because criminological research is a social, political and ethical activity. There are several roles which reflexivity can play in research, two of which can be noted here. The first concerns the assessment of validity. Validity is the extent to which conclusions drawn from a study are plausible and credible and the extent to which they can be generalized to other contexts and to other people. Validity is always relative, being dependent on the decisions which have had to be taken in the planning and conduct of research. Making such decisions explicit and, more importantly, assessing the probable effect on validity is the main purpose of a reflexive account (which is sometimes published alongside conclusions).

Second, reflection can be a form of research in its own right. This is especially the case in the critical social scientific tradition in which reflecting upon, analysing and challenging dominant ideological positions in society is central. For this reason Chapter 10, in which Barbara Hudson engages in critical reflection on types of penal policy, could easily have been placed in Part II of this volume, which is about *doing* research.

Conclusion

It has been stressed throughout this volume that the conduct of research can be expressed in terms of decision-making. Such decision-making inevitably involves trade-offs, for example trading off the weaknesses of one course of action against the strengths of another. Some decisions have to be taken about the minutiae of research, say in deciding whether to have a sample of 100 or of 120. Such technical issues matter, but so do the fundamental principles of criminological inquiry. These include validity (the pursuit of credible and plausible knowledge); politics (whose side am I on, if any?); ethics (what standards should I adopt and in relation to what?). Unfortunately, as noted earlier, the pursuit of one principle may inhibit the pursuit of another. So the most fundamental decision an investigator must take is how to position her/himself in relation to the validity, the politics and the ethics of research and the trade-offs which may have to be made between these.

SIGNPOSTS

Ethics: this refers to a set of standards which should guide decision-making about research. Sometimes these are laid down in the codes of conduct of professional bodies. These standards may apply at different stages of research, for example in the type of research problem it is appropriate to address, the ways in which data are collected and the manner in which findings are published. One fundamental principle of research ethics is that individuals, especially the subjects of the research, should not be harmed by the research. Another ethical principle often propounded is that of informed consent: that is, the subjects should be informed of the research and of its consequences and they should consent to their participation. Sometimes strict adherence to ethical principles can affect validity, for example when informing subjects about research leads them to behave in ways in which they would not normally behave.

There is an explicit discussion of the ethics of fieldwork in Chapters 12 and 13 in this Part. Chapter 2 in Part I discusses how an anticipation of ethical dilemmas should be covered in research proposals.

Politics: politics is concerned with power, with the way in which it is exercised and with what effect. Politics and criminological research are closely related because the different stakeholders in research each have interests to promote and to protect and differential levels of power with which to promote and protect them. The different stakeholders include the researchers themselves, the subjects of research, the sponsors of research and the gatekeepers. How these stakeholders relate one to another in the promotion and protection of interests can affect the progress and the outcome of research in ways which might not have been anticipated at the planning stage.

Politics and criminology connect in different ways. Chapter 10 discusses the role of critical social science as a form of political activity in its own right and Chapter 13 outlines ways in which politics can influence the course and outcome of research.

Policy-related research: this is research undertaken to help in the formulation of social policies, to evaluate the effectiveness of their introduction or to guide their future development. All research styles are comparable with policy-oriented research although it is often suggested that much more credence and legitimation is attached to quantitative research. One style of research which has had an impact on criminological research, especially in what is sometimes termed administrative criminology, is the evaluation model. This model is discussed in detail in Chapters 5, 6 and 7 in Part II. In this part, the final chapter considers policy-related research in the context of the influence of politics in general.

Critical social research: this is concerned with analysis of social phenomena in terms of structural issues, usually with an emphasis on the exercise of power in the relationship between social groupings. There is a particular interest in ideology as a means by which existing structures and social arrangements are legitimized and maintained. There is a commitment to not taking what is said for granted and also a commitment to changing the existing state of things. Critical research uses whatever methods of research will help uncover the structural and ideological bases of phenomena.

Chapter 10 provides examples of critical research such as analysis of discourses, ideology critique and standpoint research. Chapter 3 in Part II contains an analysis of the use of official statistics in a critical analysis of health and safety crimes.

Reflexivity: this is the process by which investigators reflect on their work from the early stages of problem formulation through to the publication of conclusions and their reception by different audiences. A reflexive account often forms part of a final report or book. In this a researcher makes explicit the process by which data and findings were produced, including the possible effects of the researcher on these, and thereby allows an assessment to be formed of the validity of overall conclusions. Reflexivity is also an important element of critical social research, one aspect of which concerns an examination of the role of criminology in the production of knowledge and of how this is used as a means of social control.

All of the chapters in this part provide examples of reflexivity.

QUESTIONS AND ACTIVITIES

1   As you read the following chapters write down the differing ways in which politics intrudes into social research.
2   What are the ways in which people can be harmed by criminological research? Are there some categories of people (e.g. corrupt police officers) who should not be protected against the harmful effects of criminological research?
3   What distinguishes critical research from policy-related research, if anything?
4   Write down the issues which you think should be addressed in a reflexive account.
5   After reading Chapter 11 consider the question, 'What factors can influence the ways in which criminological research findings are reported in the media?'

10

CRITICAL REFLECTION AS RESEARCH METHODOLOGY

Barbara Hudson

Contents

The need for critical criminology

Many of the questions with which criminology is concerned are best served by empirical inquiry, whether qualitative or quantitative, positivistic or inter-pretive. Criminologists engage in research to find answers to questions about why people commit crimes; why societies have higher crime rates at some times than at others; why some apparently similar societies have different crime rates; and what kinds of strategies and techniques are effective in preventing or reducing crime. These are crucial questions, and the research methods described elsewhere in this book are well-proven means of finding at least some of the answers.

There are, however, some other important questions which cannot be addressed by these means, but need the sort of approach generally understood as *critical* social science. Crime and punishment are intensely political and moral issues, and there are key questions that criminology needs to engage with which concern the moral and political stances taken towards crime and punishment by particular societies at particular times. Some examples of the sorts of question I have in mind that are of contemporary interest include the following:

- Given criminological understandings of the links between crime and relative deprivation, and between high crime rates and high degrees of social inequality, why do governments not do more to reduce social inequality?
- In view of the links demonstrated by criminologists between crime and certain forms of masculinity, why are policies such as regimes for young offenders which emphasize physical fitness, combat skills and other attributes of 'machismo' masculinity, pursued instead of penalties which offer opportunities for acquiring more caring skills and values?
- Since research has long established that reoffending rates after imprisonment are worse than for comparative community penalties, why do governments pursue 'prison works' policies?
- Crimes such as the killing of James Bulger by two young boys are, thankfully, extremely rare, so why was this incident allowed to influence policy towards young offenders in general?
- Since attempts to predict who will become criminals among cohorts of young people do not have a good success record, why are prediction studies still the form of criminological theory that is most influential among politicians and policy-makers?
- Why have theories which emphasize the basic similarity between offenders and non-offenders to a large extent been displaced by theories which emphasize differences between criminals and non-criminals? Labelling theory, for example, which insisted on the normality of 'primary deviance', influenced the emergence of policies such as cautioning and intermediate treatment in the 1980s. In the 1990s, underclass and newly respectable socio-biographical theories which demonstrate that 'offenders really are different', underpin strategies of exclusion and 'tough justice'.

These are questions of *the politics of law and order*, rather than empirical questions of causation or evaluation, and necessitate understanding the political-social context in which policy-making takes place. They also involve reflexive consideration of criminology's own role in relation to the exercise of power in society.

Practically, criminology needs to understand the political context in which it operates in order to have some idea of when it is likely to be influential, and to assess what strategies will be needed in order to get its message across. Morally, reflexive consideration is needed in order to be able to anticipate some of the likely effects of the production of knowledge about crime and crime control. For example, at a time when the politics of law and order is producing a vast increase in the numbers of people incarcerated, for longer periods, in more austere conditions, criminologies which link crime to various 'subordinated masculinities' are likely to be drawn upon rhetorically to legitimize the jailing of more men, 'particularly men of color' (Chesney-Lind and Bloom, 1997). This is a problem I feel very conscious of as a teacher of criminological theory: the most progressive and plausible theories are being developed in a social context which is all too likely to use them in vulgarized, piecemeal ways as part of a politicized construction of the criminal as 'alien other'.

Criminology is part of the apparatus of control in modern societies (Hudson, 1997: 452). Criminologists are engaged in the production of knowledge to help the work of police, probation officers, prison governors, forensic psychiatrists, community safety officers and other criminal justice practitioners, as well as politicians and policy-makers (Garland, 1985, 1988). As Foucault has shown, criminology is at the service of power; it is part of the technologies of power. The stance of the detached researcher producing 'objective' or even 'subversive' knowledge has been well described as an 'elaborate alibi to justify the exercise of power' (Cohen, 1988: 5). Of all the applied social sciences, criminology has the most dangerous relationship to power: the categories and classifications, the labels and diagnoses and the images of the criminal produced by criminologists are stigmatizing and pejorative. The strategies of control and punishment which utilize those conceptions have implications for the life-chances, for the opportunities freely to move around our cities, and for the rights and liberties, of those to whom they are applied.

Awareness of the political uses of criminological knowledge does not mean that all research into the causes of crime or the effectiveness of punishment and control strategies is to be avoided because of its possible moral and political implications; but it does mean that criminology is very much in need of a stream of critical research which exposes the political contexts in which criminological knowledge production is embedded.

Criminology, then, needs a body of research which can investigate the political/social context in which policy-oriented knowledge is produced, and which can engage in reflexive consideration of its own theoretical products (Nelken, 1994). The question now is, how to go about such reflexive consideration? Specifically, how to go about such reflexive consideration in a manner which will produce something that looks like 'proper' research, rather than the mere anecdotal and pessimistic or cynical denial of the value of any empirical criminology? Similarly how to avoid what has been termed 'left impossibilism', where any penal reform or innovation is seen as yet another instance of the widening mesh of control and repression? Then there are questions of how to select between different criminological perspectives; how to deal with media, political and professional accounts of crime and criminal justice.

The first step is to be clear about the object of inquiry. The kind of critical reflection that criminology needs to be engaged in is the investigation of the *power/knowledge* complex in which criminology has its existence. The object of investigation is the cluster of theories, policies, legislation, media treatments, roles and institutions that are concerned with crime, and with the control and punishment of crime. This material is the object of study, not part of the explanation.

The second step is to give specific, theorized, meaning to the idea of 'critical reflection'. This means choosing a theoretical perspective as carefully as one would choose a method of data collection in empirical research. Choosing between different sets of theories and concepts is equivalent to choosing between qualitative and quantitative methods, pre-coded questionnaires and semi-structured interviews. Research that is concerned with these topics is generally described as 'critical criminology'.

Within the broad heading of critical criminology there are several different approaches, some of the most influential being critical theory using the concepts and methods associated with the 'Frankfurt School' of social research; 'discourse' research using the ideas and methods associated with Michel Foucault; and standpoint research, developed particularly by feminist writers. There is considerable overlap between these perspectives, and they are mainly distinguished by the precise definition of their object of study, and by their commitment to certain political values. Frankfurt School research concentrated on culture and ideology in authoritarian societies, drawing on Marxist and psychoanalytic concepts; Foucault investigated the exercise of power in liberal societies, foregrounding the development and influence of the human/social sciences, drawing on a range of political and philosophical ideas and using a characteristically French 'snapshot' historical method; feminists have replaced the centrality of concepts such as 'class' and 'liberalism' with patriarchy and gender divisions. What the three approaches have in common is a commitment to research which contributes to the emancipation of those who are repressed by existing social and power relations.

Deconstructionist and postmodernist research of the kind that has become influential in cultural studies and other humanities fields has begun to appear in criminology (Lea, 1998), but is not yet so prominent as it is in related fields such as socio-legal studies.

Ideology critique: the legacy of the Frankfurt School

The term 'critical theory' is used first and foremost to describe the work of a group of scholars attached to the Frankfurt Institute for Social Research in the 1930s, some of whom went to the USA and elsewhere during the second world war, and some of whom stayed in Germany. After the war the Institute was reconstituted and a second generation of critical theorists, the most famous of whom is Jürgen Habermas, continued the work. Among the first generation of Frankfurt School writers, the best known among sociologists are probably Theodor Adorno, Walter Benjamin and Herbert Marcuse.

As (mainly) Jewish scholars in Nazi Germany, they not surprisingly set about investigating the rise of the Fascist state. Their studies focused on themes such as the development of the authoritarian personality (Adorno); the displacement of a liberal aesthetic in culture (Benjamin) and the rise of a politics of repression (Marcuse). These are obviously all themes which have relevance to aspects of law and order politics, especially changes from tolerant to repressive penal policies, and to changing images of offenders from the deprived to the depraved. The central problematic for the critical theorists' various inquiries are the *conditions of possibility* for the rise of authoritarian ideologies and regimes.

In some ways, the British left realists are important representatives of the tradition of critical criminology. Left realists such as Jock Young and his colleagues seek to replace ideological views of the causes of crime and the policies and practices that would have a reasonable likelihood of success in reducing crime by real – in Marxist terms, scientific – understandings (Young, 1997). The appropriate methodology is thus to start with replacing socially constructed data, such as recorded crime statistics, with data gained from victim surveys documenting people's real experiences of crime rather than accepting documents shaped by the considerations of officialdom. Theorizing can then proceed from real rather than ideologically glossed information.

The critical criminology with which I am concerned here, however, seeks not to *replace* ideologies of crime and punishment with 'real' understandings, but to study the ideologies themselves. Examples of the kind of work I have in mind are the studies of the development of 'new racism' carried out by Paul Gilroy and colleagues at the Centre for Contemporary Cultural Studies at the University of Birmingham, and analyses of the development of 'tough' tactics towards inner-city crime in the late 1980s, using paramilitary-style policing, harsher sentencing, and calls from politicians for less understanding and more punishment (Centre for Contemporary Cultural Studies, 1982; Scraton, 1987). These studies have drawn on critical theory's traditions of ideology critique, and on concepts such as populism and authoritarianism which arose in the Frankfurt School's research on Fascism and on pre-war USA, applying them to contemporary societies.

Ideology studies

These ideology investigations usually start by focusing on one particular manifestation of law and order politics which seems to signal some sort of changing ideology. In the studies of the criminalization of black youth in the late 1970s and the 1980s, Hall, Gilroy and other writers within the same perspective started by highlighting concrete examples of what appeared to be ideological innovation. In *Policing the Crisis*, Stuart Hall and his co-authors centre on the adoption of the term 'mugging' by the British press (Hall et al., 1978). They show how this term was brought across the Atlantic, used by a single newspaper and then quickly taken up by other elements of the tabloid press before appearing in broadsheet papers and more reflective television and radio programmes, and gradually entering into the discourse of politicians and criminal justice professionals, including criminologists. Gilroy and his colleagues highlight one particular document in similar fashion, in their case the police training booklet *Shades of Grey* (Gilroy, 1982). Analysis of first and second editions of this manual revealed a shift from seeing the 'black youth/crime problem' in very similar ways to the 'white youth/crime problem', that is, first as a problem of a few young troublemakers disrupting a generally law-abiding community, then moving to a view of black communities as more generally lawless and anti-authority. As happened after the appearance of stories about the involvement of Afro-Caribbean youth in 'mugging', a criminal, police-hating black community comes to be seen more widely as something that has to be dealt with in policy and in theory. Ideologies of race and criminality mean that 'black criminality' as a growing and serious problem is accorded a reality in the Durkheimian sense that it is real in its consequences for black people and for styles of policing, even if its empirical existence is open to considerable controversy.

Analysis then typically proceeds by finding other examples of similar ideological phenomena, so that the investigator may be sure that there is really something significant occurring, rather than just an isolated incident that might not be representative of any general trend. In the case of the development of ideologies of racism and criminality, further examples cited by the Centre for Contemporary Cultural Studies authors included Enoch Powell's 'rivers of blood' speech, parliamentary debates on immigration and the consequent tightening of immigration laws, and the continuing trend of criminalization of black communities as demonstrated by the introduction of the notorious 'sus' laws, later repealed but quickly replaced by new stop-and-search powers, and the development of confrontational styles of policing based on the ideas of no-go areas and disorder 'hot spots'.

This is not the place to reproduce the analysis that these researchers developed, which can be read in the works cited and others by the same and associated authors; the point is to note the method of gathering examples which together demonstrate the clear emergence of a new racist ideology. The imagery of the black man or woman in the 1970s changed from that of the hard-working, respectable family person who had come in the 1950s and 1960s in response to recruitment campaigns for labour for shortage occupations such as nursing and bus driving, to the stereotyping of second generation Afro-Caribbeans as criminal and disorderly (Pitts, 1993). An outbreak of disturbances in areas with high proportions of Afro-Caribbean residents in 1981 was interpreted in terms of this established official view of black communities as lawless, and the ideology of black criminality became further entrenched and augmented (Solomos, 1993).

Levels of analysis

Having gathered a series of examples that establish the existence of the phenomena to be understood – 'new racist' ideology which has at its core a criminalization of black communities – critical reflection attempts to explain the emergent ideology by progressing through a series of analytic levels. The first of these levels is criminological. What has criminology to say about the naming and stigmatizing of events and people, and of pejorative generalization from a few isolated events to depiction of a whole social group – Afro-Caribbeans, or 'youth' – as criminal, disorderly or anti-social? The authors of

Policing the Crisis drew first of all on Stanley Cohen's account of 'folk devils and moral panics' (Cohen, 1972). This work described the way in which bank holiday clashes between 'mods' and 'rockers' were sensationalized and generalized, and so generated a 'moral panic' about supposed new levels of anti-social behaviour among young people. The dynamics of the media treatment of a relatively minor incident at an Essex seaside resort led to massive police operations to keep mods and rockers apart. In Brighton, anyone who arrived by train in anything resembling mod or rocker clothing was prevented from leaving the station and entering the town. All of this was similar to the press handling of the first reported incident of 'mugging'. Cohen's 'folk devils and moral panics' analysis itself drew on existing criminological perspectives such as labelling and deviance amplification, and provided the first layer of theory for Hall and his co-authors.

The next step in critical reflection is to connect the criminological level with the relevant strand of wider social theory, and it is this step in particular which much empirical, mainstream criminology does not take. It is for this next, more abstract and theoretical, level of analysis that the conceptual resources of critical theory were drawn upon. Critical theory uses a basically Marxist framework for understanding ideology and the state, situating itself in the Gramscian tradition, with prominence given to ideas such as hegemonic dominance of the ideologies of the powerful. This is especially relevant for understanding the power of racist ideologies, for they work precisely through the persuasion of powerless white groups that problems such as unemployment, lack of housing, inadequate schools in inner-city areas, etc., are attributable to black immigrants straining resources by their excessive numbers and undermining traditional communities by the intrusion of their 'alien' culture. Racist ideologies may be generated by speeches made by the powerful, but they are carried and perpetuated through becoming part of the world-view of the powerless.

In the 1980s, the critical criminological project was primarily concerned with explaining the intensification of criminalization and repression of groups who were perceived by those in power as posing a threat to social and political order. The examples that were foregrounded included the use of criminal rather than civil proceedings and paramilitary policing methods against protest groups in the UK such as the striking miners and the women demonstrating against the presence of American cruise missiles at Greenham Common in Berkshire. Other examples of state repression of groups who were seen as potential threats to the social order include progressive restriction of entitlement to benefit for unemployed young people, the reactivation of ancient laws to prosecute people living on the streets and the destruction of their makeshift shelters. The 'criminalization of poverty', the 'criminalization of dissent' and the apparent shift of policing priorities from crime to disorder were examples of the new ideology which selected an ever-widening number of groups for categorization as 'the enemy within' alongside minority ethnic groups.

Again, the methodology of critical reflection was to select examples to establish the existence of an ideological phenomenon, and to enlist criminological theory and then wider social theory to help explain the emergence of the ideology and its manifestations in policy and practice. Among several criminological themes that appeared to offer some insight into the criminalization of groups as apparently diverse as strikers, women peace campaigners and the impoverished young, Steven Box's use of the *power-threat* hypothesis was useful and relevant. According to this hypothesis, state repression will be directed not at the criminal *per se*, but at subordinated groups who are perceived (accurately or not) as likely to pose a threat to existing political order and power arrangements (Box, 1983, 1987). This thesis originated in the USA, where it was formulated as an explanation of state and regional differences in black–white imprisonment rates. The rate differentials did not seem to conform to either the proportion of African-Americans or Hispanics in the populations, or the crime rates of the ethnic groups in the different states or regions, but seemed better to correlate with the level of political challenge or threat which appeared to be posed by the groups (Hawkins and Hardy, 1989). Transposed to a British context, the hypothesis seemed to provide at least the beginnings of an explanation for facts such as that the levels of police resourcing between the various counties and forces did not correspond to crime rates, but did correlate more closely with levels of disorder and unrest. According to the power-threat hypothesis, the more a dominant group perceives a potential threat to its power or well-being from subordinated groups, the more likely are three forms of discrimination:

1. restriction of political rights;
2. symbolic forms of segregation;
3. threat-oriented ideology. (Hudson, 1993: 86)

These three forms of discrimination certainly appeared to be directed against powerless and marginalized groups depicted as 'the enemy within' in the late 1980s.

   Once disorder rather than crime had been seen as the driving force of criminalization and repression in the mid-1980s, developments such as the importation of paramilitary policing methods, adopted to deal with the troubles in Northern Ireland, to mainland Britain, were easy to fit into the analysis (Jefferson, 1990). Coercive policing methods were also discussed by British left realist criminologists; indeed, the need for more even-handed and more accountable police practices has been one of the principal political arguments advanced by realist criminologists (Lea and Young, 1984; Young, 1997). Again, though, the difference between realist criminology and the strand of critical criminology being discussed here is that while realists are urging the replacement of one style of policing by another, critical criminologists (although their preferences as to policing styles would no doubt be the same as those of the realists) are engaged in analysing the ideologically driven practices of paramilitary, coercive policing and the ideologies themselves which give rise to those practices.

   Moving from the criminological level to the level of wider social theory, once again contemporary theories of ideology were developed and used which were rooted in the conceptual constructs of critical theory. Societies such as the UK and USA were, as Stuart Hall described them, 'drifting into a law and order society' (Hall, 1980), in which unemployment, homelessness, poverty and social dissent were recast as problems of crime. Instead of worrying about the problems that were faced by the unemployed, the homeless or badly housed, the poor, the addicted and the migrant, we worried about the 'crime problem' they posed for respectable citizens. Thus preventing social security fraud was given a higher priority than combating poverty; restricting 'illegal' or 'economic' immigration was given priority over assisting refugees.

Critical theory ideas such as 'authoritarianism', 'populism' and 'the exceptional state' were used to explain the growth of these repressive and criminalizing elements of 1980s ideologies. Thatcher and Reagan were, it was pointed out, engaged in populist appeals over the head of professionals, appealing directly to the same instincts of popular uncertainty and resentment that the purveyors of new racism had appealed to a decade before, so that the difficulties of mass unemployment, capital flight and inner-city decay were blamed not on the defects of rootless late capitalism, but on the very people who suffered most from these conditions. 'Scroungers', 'yobs' and 'single mothers' were blamed for the country's ills, and those who sought to support them, such as teachers and social workers, were derided and demoralized.

Hall's analysis of 'authoritarian populism' as the ideological underpinning of the Thatcher–Reagan era was augmented by studies of the dominant economic ideology of the times. Andrew Gamble's work *The Free Economy and the Strong State* (1988) struck chords with many critical criminologists. Gamble pointed out that the right-wing monetarism of the 1980s was a hands-off, *laissez-faire* ideology. Government intervention in the economy was seen as dangerous, and undesirable; the catchphrase of the times was 'Let the markets decide.' Such proscription of government action over an enormous sphere of social-political life obviously posed problems for politicians such as Thatcher who wished to be seen as strong leaders. An area of social life was therefore dramatized, and designated as the sphere where strong, vigorous political intervention was necessary, and crime was 'it'. A war on poverty would involve economic interventionism, but a war on crime involved moral interventionism and thus filled the political action vacuum perfectly. Financial deregulation of the City was accompanied by ever more rigorous and coercive regulation of the inner city (Hudson, 1995).

As the 1980s turned into the 1990s, a new element was added to this critical analysis of the war on crime by the ending of the cold war, which had implications for the social-political response to crime. Critical criminologists had already commented on the importation of paramilitary policing techniques from Northern Ireland to the mainland when a former chief of the Royal Ulster Constabulary became head of London's Metropolitan Police. With the ending of the cold war, the substitution of domestic crime and disorder for foreign and colonial struggles escalated.

Criminologists such as Nils Christie have drawn on Marcuse's analysis of the military-industrial complex to show the extent of the growth of the corrections complex (Christie, 1993). Companies such as Marconi, which traditionally manufactured military equipment, turned their production capacities to control technology such as electronic tagging equipment, surveillance equipment and so forth, while prisons and private security patrols became growth areas of employment both for those affected by the decline of traditional industries and for ex-military personnel. There is now, argue critical criminologists, an economic demand for crime, just as there has traditionally been an economic demand for periodic wars: the war on crime gives the military-industrial complex a new lease of life.

Standpoint epistemology

As well as theories of ideology, an important legacy of critical theory is the theory of knowledge which has come to be known as 'standpoint epistemology'. The Frankfurt School argued that the goal of positivistic social science – the creation of 'objective', value-free knowledge, following the methodology of the natural sciences – is unrealistic. For critical research, since standpoint is inevitable, it had better be overt. Rather than pretend to objectivity, critical theorists face squarely the 'whose side are we on' questions posed in the 1960s by sociologists such as Becker (1967) and Gouldner (1968).

Critical criminology is largely engaged with the question of the impact of ideologies and their practices on those on the downside of power relations. An important feature of critical criminology, then, is that it is linked to campaigns on behalf of the powerless, such as prisoners' rights campaigns (Sim, 1994). Critical criminology has espoused the standpoint of minority ethnic groups, of the poor and marginalized and, of course, women.

One of the most fruitful of criminological encounters has been between (mostly male) critical criminologists, and feminist criminologists who have raised issues of masculinity and male violence. Campaigning against violence against women does not signify abandonment of the standpoint of the powerless: on the contrary, one of the main arguments of feminist authors who have urged that the criminological spotlight should fall on domestic and sexual violence has been that even when such crimes are committed by people who are economically disadvantaged in relation to society in general, such people are exercising power within the crime relationship. The influence of feminist concerns has, therefore, led to reassessment of exactly who are the powerless, when we are thinking about crime (Dobash and Dobash, 1992; Hudson, 1998a; Kelly and Radford, 1987; Stanko, 1985, 1990).

Feminist criminologists have posed powerful challenges to all the streams and traditions in criminology, and there has been much debate between feminist criminologists themselves. I am not attempting to encompass all the different feminist perspectives within criminology, but merely to demonstrate that the standpoint of 'woman' has been one of the most powerful and richly developed standpoint epistemologies within critical criminology. The methodology of standpoint feminism has been described in various texts (see, for example, Cain, 1990; Harding, 1987), and been put to use in investigations of female criminality and the punishment of females. The work of Pat Carlen has been particularly important in this context since it combines clarity of discussion about method and the relationship of her work to other criminological traditions, with an illuminating empirically based understanding of the lives of women who commit crime and are entangled with the criminal justice system (Carlen, 1990, 1992, 1998).

The method of feminist standpoint criminology involves 'asking the woman question' – that is asking how patterns of crime, penal policies, crime prevention and community safety strategies, ideologies of law and order, or indeed criminological theories, affect women. It uses feminist consciousness – following research strategies that are concerned with the problems faced by women, and which are aimed to reduce oppression in women's lives. Feminist critical criminology exemplifies the traditional commitment of critical theory to acknowledging standpoints and having political/practical as well as theoretical objectives.

In recent years, postmodernism has posed a challenge not only to realist criminology, but also to this kind of critical standpoint criminology (Cohen, 1998; Lea, 1988). Feminist standpoint epistemology has been at the forefront of the debate between critical and postmodernist criminology, and has taken seriously the question of whether a single standpoint 'woman' is a valid basis for theoretical inquiry (Cain, 1990; Smart, 1990). A *rapprochement* between feminism and postmodernism can be found in the work of Daly, who has argued for the development of race–gender–class perspectives (Daly, 1993; Daly and Stephens, 1995), and by Carlen, who says that the prime referent is contingent on the question at issue (Carlen, 1998). Sometimes gender will be the starting point for an investigation; sometimes the starting point will be race or ethnicity, or poverty or some other quality.

Whilst Daly and Carlen are feminists who have responded to the challenge of postmodernism, Alison Young could perhaps be described as a feminist postmodernist. In her work she has not only 'asked the woman question' and used feminist consciousness in selecting research topics and in forming her understandings, but she has also challenged the ideological nature and the epistemological validity of emerging criminological standpoints such as that of 'victim' (Young, 1990, 1996; Young and Rush, 1994). The point of agreement between postmodernist and other critical criminologies is that critical criminology does not propose a single, unitary identity on which all research should be based, but insists that all research should acknowledge its standpoint, and that standpoint scholarship should be on the side of the oppressed in the situation in question.

The critical project in the new millennium

In the late 1990s and into the new century, critical criminologists continue to engage with the politics of law and order. Two themes have emerged as major critical projects. First, understanding the apparent reversal from the rational penal policy that sought to reserve imprisonment for the most serious offences that emerged in the late 1980s and was enacted in the 1991 Criminal Justice Act, to the 'get tough', 'prison works' policies of the 1993 Criminal Justice Act and the 1996 Crime (Sentences) Act. Second, criminologists are engaged in trying to understand the nature of the changes that seem to be taking place in the relative roles of the state, local communities and agencies, and the individual in relation to crime and community safety.

Work focusing on the first of these themes follows the familiar methodology of looking at pieces of legislation, speeches, and other ideological outputs, and decoding them within the analytic framework of state ideologies. Anthony Bottoms' account of the undermining of managerial rationalism by 'populist punitiveness' is an excellent and widely respected example of work which attempts to understand this change in legislative tone between 1991 and 1993 (Bottoms, 1995). Another good example is Richard Sparks' work on 'austere regimes' in prisons (Sparks, 1996). Work is also in process which is analysing the way in which the 'get tough' populism of Michael Howard's time as Home Secretary is being carried forward by New Labour and its equally tough Home Secretary, Jack Straw (Brownlee, 1998).

Some of the current work is also reflexive, analysing the contribution of criminological theory itself to the present politics of law and order. The American criminologist Jonathon Simon points out that victim surveys have contributed to the penal politics of the 1990s: it was not until national victimization surveys began to be undertaken, he argues, that the crime rate became an object of public discourse, and an object of criminological and policy intervention (Simon, 1996). As well as victim surveys, he says, the major contribution of criminology to penal policy (and therefore to 'get tough' penal politics) in the 1980s has been research to identify persistent offenders. This work is certainly the most influential form of criminology in the UK at present. Because identification as a likely repeat offender leads to such severe penal consequences (long periods of imprisonment, extended supervision on release), this 'prediction criminology' prompts in acute form the concern that criminology should incorporate a reflexive acknowledgement of its role in constituting the criminal as 'other' (Nelken, 1994).

Simon's work does not use the language of state ideology critique, but draws on concepts developed by Michel Foucault in his later work on governance in modern societies (Burchell et al., 1991; Foucault, 1980). Although he has commented on the 'get tough' policies of the 1990s (the paper cited above is a commentary on the 'three strikes' legislation in California), the main aim of Simon's work has been to show a reconfiguration in society's engagement with crime, away from a concern with punishing and blaming the individual, and towards managing risk and aggregate crime rates. This project of understanding changing 'master patterns' of control, the second major theme of contemporary critical criminology, is associated in Britain with the work of Cohen (1985) and, most recently, with David Garland (1996). The main thrust of the work of Cohen, Garland and Simon is to show the criminological/penal effects of the emergence of what social theorists have called the 'risk society' (Beck, 1990).

Analysing sentencing patterns of young burglars

I have drawn on this work in my own recent attempts to understand penal developments in the late 1990s, in which I have looked at the sentencing patterns of young burglars. In the 1980s, burglary was seen as a non-violent, property offence, and as such was on the non-prison track: burglars were viewed as suitable for punishment in the community. Much innovation went into developing effective community penalties for young burglars: offence-focused groups, intensive intermediate treatment and various day-centre programmes concentrated especially on gaining the confidence of magistrates and judges to use the projects as alternatives to custody. Burglary was seen as on the proportionality side of the twin track, with violent and sexual crimes on the 'risk' track, where imprisonment can be for longer than usual periods to reflect the risk of reoffending. In the latter half of the 1980s the imprisonment rate for burglary dropped by 10 per cent. Since 1993, the language of 'risk' has become ever more prevalent, and burglary has been seen less as a non-violent offence, and more as an offence with a high risk of reoffending. A consequence of this shift of penal focus is that the imprisonment rate for burglary was more than 20 per cent higher in 1996 than in 1992.

My method in attempting to understand the changes in imprisonment rates for burglary was that of critical reflection: seeking to explain a phenomenon by reference to a body of relevant theory. The point is not to establish or demonstrate that the sentencing of young burglars has changed, but to explain these changes in the context of new penal strategies. These strategies are themselves consequent upon twists and turns in the politics of law and order, twists and turns which reflect not only current political expediencies but also wider and deeper political shifts. This is what was earlier described as moving between levels of analysis. I went about this interpretation of changing sentencing patterns by starting with the figures, and moving through different explanatory levels, from official explanations through to wider social theories, looking at ideological moves from justice to risk and from inclusion to exclusion which reflect the transition from the certainties of modernism to the insecurities of late- or post-modernism. The movement between levels of analysis can be portrayed as follows:

sentencing statistics 1980–96

↓

official explanations, for example comments in Home Office Statistical Bulletins (these pinpoint the introduction of various pieces of legislation, political speeches, and events such as the Jamie Bulger killing);

↓

'insider accounts', by people who had been involved in policy-making, such as former Home Office officials and researchers (these often pointed to changing Home Secretaries, especially from Hurd to Howard, and response to campaigns by the tabloid press);

↓

accounts by criminologists which remain largely within the parameters of policy itself, sometimes known as 'administrative criminology' (these may criticize, but are not critical in the theoretical sense; they usually point to the necessity for politicians to 'act tough' without spending too much money);

↓

more reflective criminological accounts (accounts which develop middle-range concepts such as 'populist punitiveness' or 'new penology', and which are informed by broader social theory but do not themselves engage in analysis of phenomena other than crime and its control and punishment);

↓

wider social theories, which explain general movements and trends, for which crime and punishment may provide examples, but are not the whole subject matter (for example the development of social authoritarianism, risk society, the 'death of the social', the rise of communitarianism).

Conclusion

Other criminologists who are attempting to understand the changing nature of the approach to crime and the control of crime in the 1990s and 2000s, have highlighted the increasing fragmentation of powers to control and prevent crime. Writers such as Garland and Simon, and O'Malley (1996) have drawn on the body of work known as the 'history of the present' writing to demonstrate the fit between Foucault's post-*Discipline and Punish* work on governance, and contemporary developments in crime control (Garland, 1997; Hudson, 1998b; O'Malley, 1996). Garland, Sparks (1996) and myself have written about the value of this 'governmentality' history of the present literature for criminology, and have pointed to the continuing need for critical criminology to draw on wider social theory if it is to understand the politics of law and order, emerging shifts in discourses about crime and punishment, and if it is to have a reflexive conscience about its own contribution to the technologies of power. There is nothing new about all this: critical criminology is defined by its engagement with wider social theories; it is defined by its transgression of the narrow bounds of mainstream criminology which accepts the givens of official definitions of crime and official strategies for control of crime (Cohen, 1998).

Engagement with wider social theories does not mean that any off-the-peg theory will do: choice of theoretical framework depends on the project in hand, and theoretical choices will follow a kind of homoeopathic logic. Investigation of ideologies will draw on theories which have ideological critique as their object of analysis; the influence on crime control and penal strategies of changing forms of governance will draw upon social theory which is concerned with changing governmental rationalities. What is constant in critical theory is an awareness and acknowledgement of standpoints, and an explicit commitment to values of social justice and human rights.

Suggested readings

Bottoms, A. (1995) 'The philosophy and practice of sentencing', in C.M.V. Clarkson and R. Morgan (eds) *The Politics of Sentencing Reform*. Oxford: Clarendon Press.

Cohen, S. (1998) 'Intellectual scepticism and political commitment: the case of radical criminology', in P. Walton and J. Young (eds) *The New Criminology Revisited*. Basingstoke: Macmillan.

Nelken, D. (1994) 'Reflexive criminology?' in D. Nelken (ed.) *The Futures of Criminology*. London: Sage.

Sparks, R. (1997) 'Recent social theory and the study of crime and punishment', in M. Maguire, R. Morgan and R. Reiner (eds) *The Oxford Handbook of Criminology*, 2nd edition. Oxford: Clarendon Press.

Van Swaaningen, R. (1997) *Critical Criminology: Visions from Europe*. London: Sage.

References

Beck, U. (1990) *Risk Society*. London: Sage.

Becker, H. (1967) 'Whose side are we on?', *Social Problems*, 14 (3): 239–247.

Bottoms, A. (1995) 'The philosophy and practice of sentencing', in C.M.V. Clarkson and R. Morgan (eds) *The Politics of Sentencing Reform*. Oxford: Clarendon Press.

Box, S. (1983) *Power, Crime and Mystification*. London: Tavistock.

Box, S. (1987) *Recession, Crime and Punishment*. Basingstoke: Macmillan.

Brownlee, I. (1998) 'New Labour – new penology? Punitive rhetoric and the limits of managerialism in criminal justice policy', *Journal of Law and Society*, 25 (3): 313–335.

Burchell, G., Gordon, C. and Miller, P. (eds) (1991) *The Foucault Effect: Studies in Governmentality*. Hemel Hempstead: Harvester-Wheatsheaf.

Cain, M. (1990) 'Realist philosophy and standpoint epistemologies or feminist criminology as a successor science', in L. Gelsthorpe and A. Morris (eds) *Feminist Perspectives in Criminology*. Milton Keynes: Open University Press.

Carlen, P. (1990) *Alternatives to Women's Imprisonment*. Buckingham: Open University Press.

Carlen, P. (1992) 'Criminal women and criminal justice: the limits to and the potential of feminist and left realist perspectives' in R. Matthews and J. Young (eds) *Issues in Realist Criminology*. London: Sage.

Carlen, P. (1998) 'Criminology Ltd: the search for a paradigm', in P. Walton and J. Young (eds) *New Criminology Revisited*. Basingstoke: Macmillan.

Centre for Contemporary Cultural Studies (1982) *The Empire Strikes Back*. London: Hutchinson.

Chesney-Lind, M. and Bloom, B. (1997) 'Feminist criminology: thinking about women and crime', in B.D. MacLean and D. Milanovic (eds) *Thinking Critically about Crime*. Vancouver: Collective Press.

Christie, N. (1993) *Crime Control as Industry: Towards Gulags Western Style?* London: Routledge.

Cohen, S. (1972) *Folk Devils and Moral Panics*. London: Paladin.

Cohen, S. (1985) *Visions of Social Control: Crime, Punishment and Classification*. Cambridge: Polity Press.

Cohen, S. (1988) 'Criminology', in S. Cohen (ed.) *Against Criminology*. New Brunswick, NJ: Transaction Press.

Cohen, S. (1998) 'Intellectual scepticism and political commitment: the case of radical criminology', in P. Walton and J. Young (eds) *New Criminology Revisited*. Basingstoke: Macmillan.

Daly, K. (1993) 'Class–race–gender: sloganeering in search of meaning', *Social Justice*, 20 (1–2): 56–71.

Daly, K. and Stephens, D.J. (1995) 'The "dark figure" of criminology: towards a black and multi-ethnic feminist agenda for theory and research', in N.H. Rafter and F. Heidensohn (eds) *International Feminist Perspectives in Criminology*. Buckingham: Open University Press.

Dobash, R.E. and Dobash, R.P. (1992) *Women, Violence and Social Change*. London: Routledge.

Foucault, M. (1980) 'Two lectures', in M. Foucault, *Power/Knowledge: Selected Interviews and Other Writings*, ed. C. Gordon. New York: Pantheon.

Gamble, A. (1988) *The Free Economy and the Strong State: the Politics of Thatcherism*. Basingstoke: Macmillan.

Garland, D. (1985) *Punishment and Welfare: A History of Penal Strategies*. London: Gower.

Garland, D. (1988) 'British criminology before 1935', *British Journal of Criminology*, 28 (2): 131–147.

Garland, D. (1996) 'The limits of the sovereign state: strategies of crime control in contemporary society', *British Journal of Criminology*, 36: 445–471.

Garland, D. (1997) '"Governmentality" and the problem of crime: Foucault, criminology, sociology', *Theoretical Criminology*, 1 (2): 173–214.

Gilroy, P. (1982) 'Police and thieves', in Centre for Contemporary Cultural Studies, *The Empire Strikes Back*. London: Hutchinson.

Gouldner, A. (1968) 'The sociologist as partisan: sociology and the welfare state', *The American Sociologist*, May: 103–116.

Hall, S. (1980) *Drifting into a Law and Order Society*. London: Cobden Trust.

Hall, S., Crichter, C., Clarke, J., Jefferson, T. and Roberts, B. (1978) *Policing the Crisis*. London: Macmillan.

Harding, S. (ed.) (1987) *Feminism and Methodology*. Milton Keynes: Open University Press.

Hawkins, D.F. and Hardy, K.A. (1989) 'Black–white imprisonment rates: a state-by-state analysis', *Social Justice*, 16 (4): 75–93.

Hudson, B. (1993) *Penal Policy and Social Justice*. Basingstoke: Macmillan.

Hudson, B. (1995) 'Beyond proportionate punishment: difficult cases and the 1991 Criminal Justice Act', *Crime, Law and Social Change*, 22: 59–78.

Hudson, B. (1997) 'Social control', in M. Maguire, R. Morgan and R. Reiner (eds) *The Oxford Handbook of Criminology*, 2nd edition. Oxford: Oxford University Press.

Hudson, B. (1998a) 'Restorative justice: the challenge of sexual and racial violence', *Journal of Law and Society*, 25 (2): 237–256.

Hudson, B. (1998b) 'Punishment and governance', *Social and Legal Studies*, 7 (4): 581–587.

Jefferson, T. (1990) *The Case against Paramilitary Policing*. Milton Keynes: Open University Press.

Kelly, L. and Radford, J. (1987) 'The problem of men: feminist perspectives on male violence', in P. Scraton (ed.) *Law, Order and the Authoritarian State*. Milton Keynes: Open University Press.

Lea, J. (1998) 'Criminology and postmodernity', in P. Walton and J. Young (eds) *New Criminology Revisited*. Basingstoke: Macmillan.

Lea, J. and Young, J. (1984) *What is to be Done about Law and Order?* Harmondsworth: Penguin.

Nelken, D. (1994) 'Reflexive criminology?', in D. Nelken (ed.) *The Futures of Criminology*. London: Sage.

O'Malley, P. (1996) 'Post-social criminologies: some implications of current political trends for criminological theory and practice', *Current Issues in Criminal Justice*, 8: 26–38.

Pitts, J. (1993) 'Thereotyping: anti-racism, criminology and black young people', in D. Cook and B. Hudson (eds) *Racism and Criminology*. London: Sage.

Scraton, P. (ed.) (1987) *Law, Order and the Authoritarian State*. Milton Keynes: Open University Press.

Sim, J. (1994) 'The abolitionist approach: a British perspective', in A. Duff, S. Marshall, R.E. Dobash and R.P. Dobash (eds) *Penal Theory and Practice: Tradition and Innovation in Criminal Justice*. Manchester: Manchester University Press.

Simon, J. (1996) 'Criminology and the recidivist', in D. Shichor and D.K. Sechrest (eds) *Three Strikes and You're Out: Vengeance as Public Policy*. Thousand Oaks, CA: Sage.

Smart, C. (1990) 'Feminist approaches to criminology or postmodern woman meets atavistic man', in L. Gelsthorpe and A. Morris (eds) *Feminist Perspectives in Criminology*. Milton Keynes: Open University Press.

Solomos, J. (1993) 'Constructions of black criminality: racialisation and criminalisation in perspective', in D. Cook and B. Hudson (eds) *Racism and Criminology*. London: Sage.

Sparks, R. (1996) 'Penal politics and politics proper: the new austerity and contemporary British political culture'. Paper given to the Law and Society Association Annual Meetings, July, Glasgow.

Stanko, E. (1985) *Intimate Intrusions*. London: Routledge and Kegan Paul.

Stanko, E. (1990) 'When precaution is normal: a feminist critique of crime prevention', in L. Gelsthorpe and A. Morris (eds) *Feminist Perspectives in Criminology*. Milton Keynes: Open University Press.

Young, A. (1990) *Femininity in Dissent*. London: Routledge.

Young, A. (1996) *Imagining Crime*. London: Sage.

Young, J. (1997) 'Left realist criminology: radical in its analysis, realist in its policy', in M. Maguire, R. Morgan and R. Reiner (eds) *The Oxford Handbook of Criminology*, 2nd edition. Oxford: Clarendon Press.

Young, A. and Rush, P. (1994) 'The law of victimage in urbane realism: thinking through inscriptions of violence', in D. Nelken (ed.) *The Futures of Criminology*. London: Sage.

11

'LAGER LOUTS, TARTS, AND HOOLIGANS': THE CRIMINALIZATION OF YOUNG ADULTS IN A STUDY OF NEWCASTLE NIGHT LIFE

Robert G. Hollands

Contents

In the summer of 1993, I was fortunate enough to have been granted an award from the Economic and Social Research Council (ESRC) for a research project on youth culture and the city. The main aim of the research was to examine contemporary youth leisure and cultural activity, focusing particularly on the phenomenon of 'going out'[1] within the context of rapid economic change, delayed transitions out of the family household and shifting modes of consumption. Based on a small budget, and set within the colourful post-industrial city of Newcastle upon Tyne, the project sought to make a modest contribution to the academic divide between analysing youth cultures and styles and studies of youth transition in the labour market. In addition, it conveniently brought together two of my main teaching interests – urban sociology and youth studies.

I was caught completely unawares by the level of public debate, reaction and exposure the project generated, even before it officially began. News of the grant award was first reported in both Newcastle newspapers and regional television stations, before circulating more widely by appearing as a story in eight national dailies, and on Radio 5 and BBC Northern Ireland radio. It gained international coverage in Canada, through two radio interviews with the Canadian Broadcasting Company (CBC), before finding its way into the print media in Germany, Japan and Hong Kong, not to mention appearing on the front page of the *Egyptian Gazette*.

This chapter sets out to examine the ways in which both the local and national media sought to report and comment on my study of night-life cultures. In particular, it is concerned with how the project itself became 'criminalized', in terms of the way in which my research subjects were portrayed, not to mention how the study itself was drawn into an overall 'social problem' framework. The role of the local and national press and the response of Newcastle City Council and local people will be specifically analysed in this regard. Yet the chapter is also interested in the manner in which good critical research findings can act to challenge deviant and criminal stereotyping and eventually contribute to changing attitudes and inform public policy decisions.

The criminalization process: media reports and reaction to the study

Background to the research

In the autumn of 1992, I left my job in sociology at Sunderland Polytechnic to take up my present position as lecturer in the Department of Social Policy at the University of Newcastle. My new post initially entailed teaching fewer hours per week, so I set to work preparing a number of research applications. In fact I submitted five applications, all in the first six months of the job (two were jointly submitted), and to my delight, two were funded.

The application, submitted to the ESRC, the main funding body for the social sciences, was entitled 'Youth Cultures and the Use of Urban Space'. The main idea behind the project was to combine my previous work and expertise on youth (see Hollands, 1990) and leisure (see Cantelon and Hollands, 1988), with a newly developed teaching and research interest in the city and urban issues, which by its very nature contained a strong policy component. While my earlier work centred primarily around the issue of youth transitions, that is, education, training and youth labour markets, I had maintained a keen interest in the cultural analysis of youth from my days as a PhD student at the Centre for Contemporary Cultural Studies (CCCS), at the University of Birmingham.[2] Having taught in the area of youth studies for a number of years, I was also aware, from comments made by my students, that there was a lack of good studies of 'contemporary' youth cultures and subcultures, apart from those conducted back in the early 1970s (see Hall and Jefferson, 1976; Hebdige, 1979; Pearson and Mungham, 1976; Willis, 1978).

It seemed to me both innovative and logical to combine my previous interests in youth transitions and leisure with more spatial and urban concerns. One idea, which did strike me at the time, was to conduct a study of young people struggling to make the transition to adulthood on one of Newcastle's poorest estates. The 'riots' in Newcastle's west end and on the Meadowell estate in North Tyneside, had occurred only that summer and was clearly viewed as a viable topic for youth research at the time.[3] Ironically, I rejected developing such a project because I felt that it too neatly fell into the tired old 'youth and delinquency' category.

A more positive and certainly lively topic was how young adults on Tyneside were coping with blocked transition to adulthood through their engagement in youth cultures and leisure activities. In particular, one could hardly fail to notice the ritual taking over of city centre streets by young adults in most of the urban centres in the UK, and Newcastle was no exception to this general phenomenon. In fact, newspaper coverage of Newcastle's reputation as a 'lively' scene (to put it mildly) meant that it was a legitimate topic of public policy, which had implications for the city council's urban policy and economic development strategy. The evolution of a range of youth cultures and styles around 'acid house' or 'rave culture',[4] as it spread outwards from London and Manchester during the so-called 'second summer of love' in 1988 into a myriad of dance, fashion and drug styles, bumped up against more provincial youth styles and working-class traditions in 'post-industrial' urban centres.[5] It seemed to me that the city of Newcastle represented an excellent case study of how one might link up study of contemporary youth cultural styles with a cohort struggling with the transition to adulthood and hampered by high rates of unemployment and insufficient finances to leave their family household. As I was now located in a department of social policy, these were all issues of public concern which had immediate policy relevance.

These then were some practical and indeed pragmatic reasons why the research began to take on the form it did. Yet, there were also strong academic and theoretical arguments which helped frame the research question. One debate, already mentioned, was a historic and persistent division in youth research between theories of youth transitions as distinct from studies of youth cultures (for a more recent assessment of this division see Gayle, 1998).[6] In other words, with minor exceptions, there was a tendency in youth research to study transitions from education and training into work, or household transitions from a strictly 'structural' point of view, outside of any kind of cultural framework.[7] On the other hand, youth subcultural analyses, associated with the work of CCCS in the 1970s (see Hall and Jefferson, 1976 especially), had either faded out or become devoid of any transition or institutional context (like the family, community or labour market – see Hebdige, 1979). The significance of this division not only meant that there was a theoretical imbalance in youth studies – about how to bring the two areas together or how to develop different or more comprehensive theories. It also had the effect of prioritizing programmes of research at particular times. For example, there were very few relevant studies of youth subcultures in the 1980s, as unemployment and broken labour market transitions grabbed the lion's share of the attention (see Banks et al., 1992; Bates and Riseborough, 1993), with only minor consideration being given to the whole area of youth leisure and culture.[8]

Another key theoretical development of the 1980s and 1990s which occurred outside of the youth field altogether, but which was no less significant for contexting the research project in question, was a growing debate about the merits of a 'spatial' sociology (Giddens, 1984; Gregory and Urry, 1985; Saunders, 1986). With aspects of the development coming from human geography (see Harvey, 1989; Massey, 1994; Watson and Gibson, 1995) and urban sociology (Gottdiener, 1994; Savage and Warde, 1993) it was no surprise to find that there was a renewed interest in the city and urban policy issues. Many of the key ideas and concerns mooted within this developing paradigm – the changing nature of community, the relationship between city space, place and identity, notions of the divided city, the effect of global culture on localities – seemed to apply directly to a post-industrial city like Newcastle. Yet, surprisingly, there appeared to be very little written at the time about young people and cities, despite their obvious colonization of elements of urban space.[9]

Armed with a theoretical justification, as well as an empirically obvious and practical policy-based rationale for studying youth cultures in the city, I prepared and submitted a bid to the ESRC in the winter of 1992. It is instructive to quote from the aims and objectives of the application itself, to demonstrate the original intentions of the research, before moving on to show how it was presented and commented on by the media, the general public and local government.

The main aim of the research, as expressed in the application, was 'to explore and document the variety of cultural meanings young adults attach to their evening use of urban space'. This area of study was to be set within the wider context of rapid economic restructuring and change, delayed transitions out of the family household and shifting modes of youth consumption and culture in the 1990s. Ironically, a second related aim was to understand how young people's own 'cultural experiences *contrast* with public and media produced conceptions of contemporary youth cultures' (original emphasis). Little did I know how significant this contrast would become for the research. The proposed methodology to be utilized was a well-respected tradition of ethnography and participant observation which it was hoped would extend 'some of the political and policy implications of this approach'. Again, it is interesting that both of these emphasized points – the methods used and the policy relevance – became major features of the public and media reaction towards and indeed criticism of the study. While part of the reaction may have had to do with the fact that participant observation methods are often seen as producing 'soft' data (rather than 'hard facts'), much of the criticism, I felt, came from a general misunderstanding of what sociologists actually do when they conduct research.

Local media and public reaction to the research

I received notification of the granting of the award in the summer of 1993 and the university immediately included my name, research title and amount of the award on a database available to the public. Unaware of this procedure, I was surprised to receive a telephone call in late August from a journalist working for the *Journal*, a local paper. I spoke with him at length about the research, explaining both the background and the rationale behind the study. Believing that it would be a local interest story, I agreed to pose for a photograph (perhaps an unwise decision in hindsight), both outside and inside a local bar in the city centre. When a picture of me holding a drink inside the bar was later sold to the national and international press and embroidered with racy headlines and bylines, it was easy to forget that the contents of the glass I was holding was actually Perrier with a slice of lemon.

The original story printed by the *Journal* (reproduced in Figure 11.1) on 27 August 1993 was one of the better and more accurate depictions of the aims and objectives of the research. This was partly because the journalist in question was relying on the researcher for information rather than writing the story first, and partly to do with the newspaper's reputation as a quality local broadsheet. However, while it contained numerous direct quotes from myself and covered much of what the research was about, there was perhaps undue stress on a particular part of the city centre called the Bigg Market,[10] focusing specifically on the question of why so many people flock there.

Part of the reason for this focus may be due to what media analysts call an 'inferential structure' (Hall et al., 1978). What is meant by this term is that previous coverage of a phenomenon, in this case Newcastle night-life, is used to 'frame' and contour any future stories on a particular topic. If Newcastle night-life is synonymous with the Bigg Market, and this area already has a reputation (for violent, rowdy, drunken, promiscuous behaviour), then the reporting of any stories related to this topic will tend to be viewed in this light. For instance, while the text of the story mentions other areas of the city and raises a range of important issues – like changing gender relations, student–local conflict, access for different groups, the economic/domestic and cultural context of the research – the headlines, photos and opening text disproportionately help set the agenda for the piece and subsequent coverage (that is, the research is primarily about this area of the city and why people go there). The article also ends on an unhelpful note, by hinting at the fact that there may be a simple, common-sense explanation for the question that the media itself has constructed. The scenario for much of the media coverage to follow was set by the following remark, found on the adjacent editorial page:

The *Journal* (at 28p) is happy to save the Economic and Social Research Council some £15,999.72p – and the lecturer's time and trouble – by putting forward our own revolutionary conclusion. Could the Bigg Market's enduring popularity be something to do with the fact that numerous people in the North-East simply like going out and enjoying themselves?

The other local paper, the *Evening Chronicle*, printed a story on the same day in their morning edition, without actually having talked to the researcher at all. Here the research was reported within the context of a 'political storm', with the study being criticized as 'absurd', 'common-sense' and a 'waste of money'.

Figure 11.1   *Bigg night out*

The main thrust of the criticism was that the project was a waste of public money, which could have somehow been spent on other council services facing cutbacks.[11] The article contained some seething quotes from the opposition Liberal Democrat leader, and a number of critical comments from a supposedly 'randomly' selected group of young people themselves. The evening edition carried much the same story, expanding the range of comments from councillors and, due to my insistence and initiative, included some quotes in defence of the research. While the most cautious comment came from the then Labour leader of Newcastle City Council, Jeremy Beecham, suggesting that he was doubtful the council would learn anything new from the research, the most significant commentary, in terms of this newspaper's argument, came from the two opposition leaders. The Liberal Democrat leader, not having read the research proposal, argued that the real issue was one of disorder, drunkenness and safety. The Conservative leader went even further down this criminalization road by suggesting that the whole idea was a 'crackpot' project and that the money could be best spent on 'an extra closed circuit television camera, for example'.[12]

The same day, the story was picked up on by the local television media and I conducted one taped and one live interview. While the gist of both reports was broadly similar, continuing the themes of 'common-sense' and 'waste of money', they too contained elements which continued to construct the research and the research subjects within a 'deviance framework'. For instance, the Tyne Tees Television report immediately linked the area of the Bigg Market with the 'fat slags' – two hard-drinking, promiscuous, Geordie women comic strip characters from the popular magazine *Viz*.[13] The feature also carried a number of interviews with young people themselves, one of which replicated many of the deviant stereotypes associated with 'nights out' in the area in question. Consider the following conversation between two young lads (around 15 years old), when they were asked by a TV reporter why they thought young people came down to this part of the city:

> Lad 1: It's just the main place to be in town. You meet a whole lotta people down here. The only thing bad about it is the fighting. There's a lotta fighting.
> Lad 2: Any amount of birds [women] though. Good-looking ones and that. That's why single young men like us come down here.

The report on the local BBC station followed a similar line of argument and format. One young woman who was interviewed was quoted as saying: 'You can just run riot and do whatever you want', when asked why the area was so special. Again there was reference to the Bigg Market being 'renowned for its night-time high spirits', accompanied by a number of clips of drunken, aggressive groups of people leaving pubs in that area at closing time.

What was so dominant in all this local reaction to the subject of the research was the degree to which the activity was unquestionably framed with a 'social problem' perspective (Goode and Ben-Yehuda, 1994; Rubington and Weinberg, 1995). So much so, that public figures felt comfortable enough to pontificate about the research without either reading the proposal or stopping to realize that the very rationale of the study was actually to provide empirical evidence for assessing some of these 'common-sense' claims. For example, no one questioned during this local coverage provided any real evidence for their assertions. Curiously, the local media worked in such a way as to appear to have co-opted young people themselves into reproducing criminal and deviant discourses in relation to night-life activity.[14]

National newspaper coverage of the research

National newspaper coverage of the research began immediately after the local press stories hit the streets. While the principal researcher was able to speak to a number of journalists, the vast majority intimated that they had already written the story and were simply looking to confirm details or get a quote. In talking with numerous journalists, I began to realize that the vast majority were relatively uninterested in the research itself, but rather were intent on pursuing it as either a quirky human interest piece or as a 'social problem' type story. Attempts were made to ask me about my own drinking and night-life habits, and when information was not forthcoming, some papers simply made up quotations and attributed them to me.

The main thrust of the national newspaper coverage reproduced some of the local criticisms of the research and furthered images of Newcastle night-life as both dangerous and deviant. For example, the Bigg Market was described in the *Independent* (28 August 1993) as a 'notorious street full of pubs and theme bars' and 'the social equivalent of the D-Day landings' on a Friday and Saturday night. The clientele was further stereotyped as 'bearing a remarkable resemblance to Gazza' (the infamous Tyneside-born footballer), and were typified as 'massing in the street, drinking and occasionally beating one another up'.[15] The *Daily Express* (28 August 1993) suggested I would be joining revellers in an area 'which has a reputation for drink-fuelled violence'. Curiously, none of the national reports mentioned that a key element of the research was to empirically test the hypothesis that violence levels were higher here in comparison to certain other parts of Newcastle city centre.

While some of the national coverage included local criticism of the research in financial terms, inflating the issue to a country-wide level (i.e. the *Daily Mirror*'s headline '£16,000 nights are on us!'), interestingly enough the vast majority of stories took a 'human interest' angle. However, in doing so they also, perhaps inadvertently, portrayed the research topic, not to mention the researcher himself, as somewhat deviant for wanting to study such a topic. Even the quality broadsheets like *The Times, Daily Telegraph* and *Independent* all began their stories with a variation on the line 'A university lecturer has been awarded a £16,000 grant to go out at weekends and drink' (*Independent*, 28 August 1993). The researcher was colourfully described in various stories as a 'young 37', 'a fun-loving academic' and a 'wine buff'. The *Today* newspaper headline of '£16,000 grant to go on a pub crawl', was followed up by a set of quotes allegedly attributed to myself, and contained a section entitled 'Bob to study wine, song and women'. It also contained the following: 'It means the poor chap will have to down pint after pint, night after night, while watching and noting the behaviour of fellow tipplers.' The most sensational headline however, was that of the *Daily Star* (28 August 1993): 'Boozy Bob's Going On a £16,000 Pub Crawl'.

In summary, while the national press chose to highlight the more quirky aspects of the research topic, they too reproduced images of my research subjects as deviants and law-breakers. What had started out as a positive project, looking at how young adults were creatively coping with a relatively difficult transition to adulthood by seeking their identity elsewhere, was reported as a somewhat dubious study looking at what was an 'obvious' social problem. The next section of this chapter looks at the ways in which some of the research findings challenged and indeed shifted public opinion of the meaning and significance of night-life activity away from a purely 'deviance-based' paradigm.

Decriminalizing the research topic and subjects: the role of research findings

One of the key contributions made by sociological research is to help society question some of its deepest assumptions about everyday phenomena. Sociological work involves invoking the 'sociological imagination' (Mills, 1975) – requiring us to 'think ourselves away from the familiar routines of our daily lives in order to look at them anew' (Giddens, 1997: 3). This does not imply that sociology is completely divorced from 'common-sense' ideas – indeed sometimes it contributes to them – only that it should always subject the 'obvious' to detailed scrutiny and critique as a matter of course.

This indeed was one of the main objectives of my research on young adults and Newcastle night-life. Looking at the wider economic, social and cultural context, it seemed clear to me that there was much more going on in this sphere of young people's lives than popular representations revealed. Ironically, one of the positive features of all the press reaction and attempts to criminalize the research process and research subjects was that it provided a ready-made critical foil for the study before it had even begun.

Part of the rebuttal came at the very moment the research was under attack. In various television, newspaper and radio interviews, and letters to the editor, I sought both to defend the research programme as well as to question many of the unsubstantiated views and ideas that were being asserted by the press and members of the public. In the television interviews I challenged the assertion that the study of popular culture was 'trivial' and 'common-sense', pointing out that the research would indeed provide a different interpretation of night-life culture, not to mention produce some very concrete policy proposals. I also asserted that the very group in question, young Tynesiders, were rarely given a voice to express their views on their own activities.[16] Instead, interpretation of this is often left either to people in positions of power, or to those who stand outside the culture looking in, and tends to be patronizing in tone. Finally, I also questioned what evidence existed for some of the assertions about the incidence of violence, vandalism and promiscuity on nights out, claiming that the research would indeed finally provide some solid social scientific information on these matters.

The publication of the research, first as a 45,000-word report and then as a book entitled *Friday Night, Saturday Night: Youth Cultural Identification in the Post-Industrial City* (Hollands, 1995) was the main vehicle for both publicizing the research findings and challenging some of the main stereotypes of night-life activity.[17] Curiously, despite the issue of a press release, the national media in particular appeared less interested in the research than one might have expected. However, there was again good local coverage and the report was later more widely reported in the *Times Higher Education Supplement* and in a popular dance magazine, *Mixmag*.

The core finding of the research was that the main social meaning of 'going out' was not confined simply to the consumption of alcohol or the search for a sexual partner, but was more closely connected to young adults' desire to construct a sense of local identity, and indeed 'community', through an extended leisure-based ritual involving socializing with peers. As the press release starkly put it:

> Contrary to popular opinion, nearly twice as many young adults go to pubs and clubs to socialize with friends, compared with those who say their primary intention is to get drunk, and going out to dance and listen to music is more highly rated than meeting someone of the opposite sex.

The shift in the social meaning of 'going out', from a simple 'rite of passage' into adulthood to a more permanent socializing ritual involving the search for identity and belonging, was put down to broader shifts in economic, domestic and cultural life. Again, this point was expressed in the press release:

> Many young adults are turning towards consumption styles and activities in the city to express themselves in place of the loss and uncertainty surrounding traditional sources of identity such as work, education and marriage. Pride in being a 'Geordie' for many young people has less to do with their jobs or industrial production, and much more to do with celebrating a version of this local identity out in city centre streets and pubs.

This main finding of course was not suggesting that heavy drinking and sexual display and symbolism were unimportant elements of night-life activity. Alcohol remains a potent western cultural icon for sociability and the construction of temporal and spatial communities – the role it plays in wedding celebrations and funerals, for instance. And while alcohol consumption on nights out was quite high, unhelpful stereotypes about the drinking habits of Bigg Market regulars were put into context when contrasted with the much higher weekly consumption rates of non-local university students. Furthermore, the vast majority of local respondents reported that they suffered very few problems related to alcohol either during or after its use.[18]

The same argument can be made in relation to public perceptions of sexual activity and promiscuity on nights out. Again, the research findings did not deny that sexual symbolism and display is one of the most 'visible' features of night-life culture. Yet, it was only a minority of people who cited this as the 'main' reason why they went out on nights out. The research also put into context the often wide disjunction between wishful thinking and the reality of having sexual relations with someone as a result of a night out. Indeed, nearly half of the sample believed that much of the talk about meeting and sleeping with someone on a night out was somewhat exaggerated. Rather than rampant promiscuity, it was found that the figures more accurately reflected national patterns of sexual activity for this age group, the average number of sexual partners met on a night out in the last year standing at a modest 1.5 per person.[19] The familiar pattern of a small number of people having a high number of sexual partners, and a large majority having none at all or simply being in a steady relationship, was also evident. Local women in particular, rather than representing the comic-book stereotype of the 'fat slags', had in fact the lowest number of sexual partners met on a night out of all groups studied, including student women.

The most significant findings regarding challenging the crime and deviance framework, however, were those to do with the incidence of violence and vandalism engaged in and observed on a night out. Many English city centres, typified as filled with drunken youngsters on the rampage are often seen as places of extreme violence. Indeed, crime rates in metropolitan areas and city centre bars and pubs often exceed those of non-metropolitan areas and the surrounding suburbs (Hope, 1983). However, it is important to remember that there is a much higher density of people congregating in city centres at the weekends, and that violent crime is a very small proportion of the general crime figures associated with urban areas. (Assaults and woundings formed only 3 per cent of all crimes committed in Newcastle city centre when the research was conducted.)

Public perceptions of the city as a dangerous place are more likely to be held by those who do not frequent such areas. An overwhelming 82 per cent of our sample of young adults felt that the danger element of going out was exaggerated. And even for those who felt there might have been particular 'trouble spots' in the city, most admitted that their feeling was motivated more by hearsay than actual experience. Nearly 80 per cent of the sample had never been involved in a violent incident – and this involved some 22,000 visits to Newcastle city centre in total. For those unlucky enough to have been subjected to some form of violent attack, the average number of incidents recalled was 1.6 altercations over the entire time they had been going out. Bigg Market regulars curiously were slightly less likely to have been involved in a fight, although they accounted for a disproportionate number of incidents (that is, a small number of them were engaged in a high number of incidents). The most telling statistic was that when one compared the incidence of fights to the sum total of nights out, the ratio was one altercation to 970 visits, or roughly one fight experienced every nine and a half years going out.

Of course, one of the explanations of the perception of the city as a dangerous place is derived not from experiencing violence directly, but from witnessing it. Approximately three-quarters of the sample had seen fights take place in the city centre. At the same time, the average number of incidents 'ever witnessed' per person was around a dozen. Again calculating against the number of visits made, this meant that the average number of fights recalled per year was around three. Surprisingly, that section of the local sample who were more likely to frequent the Bigg Market area of the city, actually reported witnessing less violence than other areas. This view was supported by an independent six-month study of incidents recorded by CCTV cameras, which demonstrated that violence was no more prevalent in the Bigg Market area than in other parts of the city (Centre for Research on Crime, Policing and the Community, 1993). So another of the media stereotypes about this area of the city dies hard when stacked up against some real evidence.

Involvement in and the witnessing of acts of vandalism were even more infrequent. Only 13 per cent mentioned ever being involved in such an activity and most examples cited were petty in nature. University students, in fact, were more likely than locals to admit an act of vandalism. A higher percentage (around 50 per cent) had 'ever witnessed' an act of vandalism on a night out, but again one needs to place the number of incidents against the high number of times people frequent the city centre.

Ironically, the research uncovered a number of issues not 'normally' associated with the field of crime and deviance, making them all the more disturbing. For example, the incidence of sexual harassment experienced by young women in the city far exceeded any acts of violence and vandalism: 60 per cent of our total female sample had experienced some form of sexual harassment, with half experiencing actual physical contact. Local women were especially vulnerable, with nearly three-quarters being affected. In addition, numerous incidents of racism were recounted. Furthermore, the research showed that 40 per cent of the sample felt that the night-life facilities were not easily accessible to ethnic minorities. Three-quarters also either did not know or did not feel that Newcastle night-life catered enough for the needs of the city's gay population. Finally, many of the respondents were acutely aware of tensions, conflicts and divisions between the local and the university student population.

Curiously enough, very few of these latter issues received any newspaper coverage when the report and book were press-released. This is interesting in the sense that it reveals that the media and the public often have very limited definitions of what constitutes deviant and criminal behaviour in cities. What subsequent coverage did achieve was twofold. First, media reports of the research called into question some of the unfounded assumptions about the incidence of crime and deviance surrounding night-life activity. Second, by focusing on some of the positive aspects of nights out (i.e. identity, community, economic benefits to the region, service satisfaction, policy suggestions to improve provision, etc.), the research shifted some of the ideological ground about how politicians, the media and even the local population had historically looked at night-life culture as a social problem, requiring containment and repressive legislation.

For example, the *Journal* – the paper that released the initial story about the research that created such a media frenzy – printed an upbeat story on the main findings of the report and the need for Newcastle to develop a more diverse 'café culture'. It also mentioned both the low number of sexual partners met on a night out and highlighted that the violent reputation of the city was unjustified (*The Journal*, 6 May 1995). Similarly, the *Northern Echo* (10 May 1995), a regional newspaper, contained an article about the project report which used the phrase 'there's more to it than sex, violence and alcohol'. Besides reiterating some of the myths about the incidence of sex and violence, the article also gave some space to the issue of sexual harassment and questioned the attitudes of local politicians who saw youth in the city as a social problem rather than an economic resource. Finally, a local Tyne Tees Television report on the research began its feature by stating that the research showed that 'binge drinking and fighting are not typical', focusing exclusively on the more positive aspects of going out (identity, sociability, friendship), through looking at the experiences of a group of young women.

One of the most comprehensive reviews of the research and its questioning of the criminalization process was a piece in a local entertainment listing magazine *Paint it Red*. It suggested early on in the article that the report contained information which 'calls into question much of the hysteria surrounding young people and night-life' and 'does not further portray the young as pill popping boozing hooligans who go out simply for a fight' (see Figure 11.2 for a copy of the full article). Furthermore it contained a lengthy quote from myself which called into question images of deviant youth:

> there's been many moral panics regarding the young but the social problems we discovered weren't the predictable ones. It's very easy to focus on violence, but the chances of attack in the city are very low. In fact, focusing on violence detracts from other social problems we discovered to be prevalent like racism, homophobia, sexual harassment and local/student conflict. ('Bar room blitz', *Paint it Red*, 89, March 1995: 11)

The main concern of the rest of the article was with the positive benefits of a vibrant night-life and the need for local politicians to take up some of the more imaginative and creative policies discussed in the report.

Considering the media frenzy at the beginning of the research process, national coverage of the results was disappointing and hard evidence of the phenomenon was obviously considered less newsworthy. Despite a press release issued to all the national papers, only two sources (neither involved the first time around) picked up the story. Perhaps this was partly to do with the fact that there was a research report about young people which had something positive to say – not a 'news value' normally associated with youth (Griffin, 1997). However, both national sources emphasized the positive and creative potential of this activity/age group. The *Times Higher* (26 May 1995) said that the report 'should lead to improvements in the existing night-life, rather than opposing its extension on the assumption that more night spots will require more policing'. The other response, from the national dance magazine *Mixmag* (2 (51), August 1995), was entitled 'Kids prefer clubbing to sex', and focused almost entirely on the question of greater investment and improvement to existing night-life services, and the energy and talent that exists in urban youth cultures.

Public policy effects

Informing and indeed educating the public via the media around a social issue is one thing. Yet all researchers conducting work with a social or public policy

BAR-ROOM BLITZ

Dr Robert Hollands' controversial £16,000 survey into the North East's drinking habits has just been completed. It makes for fascinating reading, as Jon Bennett discovered.

Many of you may have seen the hysterical press coverage surrounding the announcement that Newcastle University's Dr Robert Hollands had been awarded £16,000 to conduct a sociological research into youth culture in Newcastle. Headlines like The Star's 'Boozy Bob's going on a £16,000 pub crawl' and The Chronicle's 'A Bigg Mistake' were scornful of his suggestion that the phenomenon of 'going out' was worthy of study. However, he has now completed the report and it is a fascinating dissection of how nightlife operates in Newcastle. He admitted that the negative attention had been somewhat dismaying. "The press coverage was really surprising. I thought it would be a small local story and then it's not only in the national papers but is in Der Spiegel in Germany. It was a laugh for them but now there is a report, the most extensive of its kind, which needs to be taken seriously. There is information in the report that calls into question much of the hysteria that surrounds young people and their nightlife. This will hopefully alter people's perceptions, and the stigmas and the policies and ideas that come from young people themselves will be discussed.

Holland's report focuses upon the development of what he calls, "Britain's nightclub culture" and the effects this has had. It does not further portray the young as pill popping boozing hooligans who go out simply for a fight. Instead it is a thorough investigation into the way in which Newcastle's nightlife has developed. It looks at behavioural patterns, class conflict, crime and other issues but the tone is not the usual scolding one of the moral guardian referring to the iniquities of the young.

"That is, of course, the popular misconception but it is an image that has been fostered by the government and the media. There's been many moral panics regarding the young but the social problems we discovered weren't the predictable ones. It's very easy to focus on violence, but the chances of attack in the city are very low. In fact, focusing on violence detracts from the other social problems we discovered to be prevalent like racism, homophobia, sexual harassment and local/student conflict, issues that it seems nobody wants to talk about but are undoubtedly present.

"Its been assumed that the reason for going out is obvious, but our research shows that it's not just about pulling and getting drunk. There are new trends such as ladies only nights, which reflect women's changing position in the economy, giving them freedom to use the city more. What

Way–hey!

struck us most was the solidarity of these people who go out in a group. They'll go through conflict with spouse or family just to have a night out."

Hollands is extremely persuasive when he stresses the effect that economic developments have had on British society. "There has been an obvious change caused during the economic decline of the 80's and 90's, but if people can't work they will still go out. Acid house and rave were symptomatic of this. People are still looking for work, but because of unemployment and higher education the transition between what typically constitutes youth and adult has lengthened unbelievably and again this is hugely important. People are behaving in a youthful manner, being boisterous for a far longer period now than before. Essentially therefore society is changing."

It is when Dr Hollands turns to policy suggestions for the future that the report is at its most illuminating. "Typically a local Conservative councillor argued that for the £16,000 spent on the report another security camera could have been bought. That comment was made without any knowledge of violence in the city and was simply a knee jerk reaction about control. That has been the policy regarding nightlife in Newcastle, controlling nightlife, yet despite this total lack of encouragement youth culture has flourished. Therefore who knows what could happen if it was helped? Essentially the report is attempting to look at these issues in a positive way. We discovered that there's a lot of culture in Newcastle, it just goes home at 11 o'clock. Youth culture and the night time economy encompasses a lot of products and could be the basis for revitalising the local economy. It's not just about people working in the service industries, it's the manufacturing that surrounds the nightlife culture. For example, clothes, music and breweries, and in the report we've attempted to give a platform for this and express what young people told us they want.

"Young people are saying they want extended licenses, a diverse club scene, cafes. It's pointless reading the report to be astonished that two thirds of young people have taken drugs on a night out. We have to focus on the positive. For example, look at how other cities such as Manchester and Dublin are using youth culture as a positive force for economic re-generation. We can learn from them.

"The city council has a role to play and while their response would be to say that they're already cutting basic social services, they do have an important role. There has to be a vision for the city or nightlife will suffer. That was the point of the report, to show what's happening, collect some ideas and start the debate."

Dr Hollands is currently in negotiation with publishers and the report should be available in the next few months. If you too are sick of people moaning that Newcastle has nothing going on then I highly recommend it. It is the most extensive study of this city's culture ever and as such should be obligatory reading. At the very least it rubbishes the stigmas that go along with being a member of the youth of today.

11

Figure 11.2 *Bar-Room blitz*

Figure 11.3   *Public policy effects of research*

focus ultimately hope that policy-makers will both hear of and take on board at least some of their recommendations. The actual public policy effect of research is often harder to trace or measure accurately. However, despite Newcastle City Council's initial coolness to the topic of the research project, the report was welcomed and praised by both a number of politicians and officers. I was asked to present my findings to the council's '24 Hour City Working Group' and was offered membership of the (now defunct) committee. While there remained some opposition to the issue of extending pub licensing hours upon publication of the report (see the *Evening Chronicle* report 'Late drinks a Bigg mistake', 31 August 1995, Figure 11.3), a year later saw almost a complete reversal of the council's position (see the *Evening Chronicle* article on 5 September 1996), 'Christmas cheer for pubs and clubs', Figure 11.4). By 1999, articles about relaxing pavement licences and creating a more continental atmosphere are much more commonplace (see also the *Herald and Post* piece, 'City goes al fresco', 7 April 1999, in Figure 11.5).

While Newcastle City Council has proceeded slowly and cautiously regarding the licensing issue, offering extended and new licences to a select number of establishments (not to mention relaxing hours on New Year's Eve recently), there has also been a slow growth in establishments characterized under the general banner of 'café culture'. But perhaps the most significant change has been the shift in language away from a purely social problem perspective on youth in the city, to one which recognizes that young adults provide a great deal of the energy and creativity in the development of a safe and economically viable night-time economy. While discourses of 'louts, tarts and hooligans' will no doubt continue to surface and influence some people's

EVENING CHRONICLE, Thursday, September 5, 1996   ● LOCAL NEWS ●

Christmas cheer for pubs and clubs

● STRICT RULES –
Tim Hibbert

NEWCASTLE'S pubs and clubs may be allowed to stay open for another hour's drinking in time for Christmas.

City council licensing chiefs are expected to back plans by the Home Office to allow licensees to stay open later on Fridays and Saturdays.

This would mean the pubs could continue serving until midnight and clubs until 3am. But the move has already

By PETER YOUNG
Political Editor

sparked off fears of an increase in drunkenness and disorderly behaviour.

And council chiefs are warning that if the proposals are adopted, drinkers in Newcastle will be on trial and late licences would be revoked if there was trouble.

The new regulations are

expected to come into force by the end of the year and the Home Office is consulting local licensing chiefs on how they should be implemented.

The proposals will be discussed by councillors at a meeting of Newcastle's public health and environmental protection committee on Monday.

Tim Hibbert, head of public health and environmental protection, is suggesting the council backs pub opening until

midnight, but with the condition that 11pm closing will be reimposed if there is trouble.

The local licensing justices could do this by imposing a restriction order when they grant an extension.

There will be consultations with the licensing trade, but Mr Hibbert says the council is ready to consider change with the minimum of bureaucracy as long as the public is safeguarded.

Figure 11.4   *Christmas cheer for pubs and clubs*

HERALD AND POST   www.herald-and-post.co.uk   **April 7 1999**

City goes al fresco

By Adam Murray

MOVES to turn Newcastle into a continental-style cafe city have stepped up a gear with the first venues receiving the new pavement licences.

A flood of applications are now set to hit licencing chiefs, with more and more venues keen to serve customers at tables outside their establishments.

Council bosses started the ball rolling by granting three pavement licences to popular watering holes Blake's, in Grey Street, and Cafe Neon and Chambers, in the Bigg Market.

Continental

Coun Tony Flynn said: "We want to give the city a more continental flavour and make it a pleasant place for tourists and visitors as well as everyone who lives and works here.

"There will be the opportunity for more pavement licences in the future, especially in the historic Grainger Town area, and we feel it's all part of taking Newcastle into the next Millennium."

Dave Hattam, owner of Blake's which has been open seven years, received licence number 0001 and was delighted that New-

● OPPORTUNITY: Coun Flynn

castle had now come into line with other Northern cities.

He said: "This is something we've wanted to do for the last two years so I'm delighted it's finally happened.

"As well as bringing in extra business, it will give a continental feel to the city centre which I think is really lacking."

The scheme is being run by the planning and transport division at the civic centre, with officers happy to advise proprietors on preparing their application.

Coun Flynn added: "We are expecting quite a few applications, but I must stress that each application is looked carefully and there are vigorous guidelines - especially on preventing unnecessary obstruction."

Figure 11.5   *City goes al fresco*

views of city life, there is now a much wider range of images and scenarios to compete with such limited deviant stereotypes.

Conclusion

At first glance it may appear a little odd to include a chapter on a topic such as night-life culture in a book about researching crime. However, what this chapter demonstrates is that the criminalization of research and its subjects can occur in quite diverse fields, and studies of youth, leisure activity and the city are no exception. Indeed it is not unusual for the media and politicians to make the youth-culture social problem connection. For this reason, youth researchers in particular need to be careful not to unproblematically reproduce 'negative' representations of youth 'as trouble' (Griffin, 1997; MacDonald et al., 1993).

This is clearly a problem for youth criminologists: for them young people, almost by definition, are constructed through a deviance-based framework. Yet the criminalization process can also occur in seemingly more neutral topics like studies of youth leisure and cultural activity. The combination of elements like alcohol, drugs, sex, urbanism and violence/vandalism, in conjunction with the term 'youth', often work to literally create a social problem (when in fact there isn't one) and can help construct new forms of youthful 'moral panics'. While this situation is partly helped along by the role of those powerful in society – 'primary definers' such as politicians, the police and other 'moral entrepreneurs' – the media also play their part. As mentioned, concern with negative news values and the importance of inferential structures (previous ways of reporting on a phenomenon) act to limit and contour what it is possible to say about a social phenomenon and who can speak. As the late E.P. Thompson (1980: 5) so eloquently put it:

> It is not only the number of 'responsible' views which are determined by the media. They also determine, to a great degree, the questions which it is possible to have views about and the form in which those questions arise . . . Thus even those who initiate controversy lose control over its course at the moment when the media takes it up.

This is not to argue, however, that sociologists and criminologists should tactically avoid the media machine, conduct their work in relative obscurity or pick only safe or boring social sciences topics to study. As the sociologist Anthony Giddens (1990) has persuasively argued, sociological knowledge can, and indeed should, circulate amongst and seek to influence public discourse on the social issues of our times. The fact that we are never completely in control of that process does not absolve the sociologist from critically engaging in work that is inherently controversial.

This chapter has suggested that despite some inherent difficulties in studying the issue of young people and night-life, and in spite of media attempts to criminalize the research subjects and process, good quality research on such a topic can influence public images and stereotypes, not to mention public policy. In this particular study, data collected from young adults themselves called into question media assumptions about the incidence levels of violence and vandalism experienced on nights out. It also sought to locate and contextualize behaviours such as alcohol consumption and sexual activity within a more sociological framework, as well as measuring their actual incidence, thereby moving the debate about these behaviours and symbols outside of a strictly deviance-based framework. At the same time, the research in question actually revealed a number of other social issues which had remained buried under the weight of criminal and deviant discourses. For instance, problems like sexual harassment of women in the city, the effect of racism and homophobia and conflict between students and local residents in the area. Other issues related to improving the quality and variety of night-life services and policies for expanding the night-time economy were also raised. Ironically, policies to improve and expand night-life, including extending licensing, creating alternative forms of provision linked to more of a 'café culture', better transport services and policing, and increased safety, might actually work to reduce the incidence and risk of crime in the city.

Finally, any good research, whether or not it has to do with crime and deviance directly or even inadvertently, should seek to shift some of the ideological ground and common-sense assumptions surrounding the phenomenon in question. In this case, it is clear that there has been a reordering of public discourses about night-life away from a purely 'criminalized' or 'social problem' framework, towards a more positive position concerned with energy, creativity and opportunities for developing the night-time economy. And although aspects of night-life culture may never be completely divorced from outdated arguments of 'lager louts, tarts and hooligans', there is now a range of more positive and competing images to build on.

Suggested readings

Griffin, C. (1997) 'Representations of youth', in J. Roche and S. Tucker (eds) *Youth in Society*. London: Sage.
Hammersley, M. (1992) *What's Wrong with Ethnography?* London: Routledge.
Harvey, L. (1990) *Critical Social Research*. London: Unwin-Hyman.
Hollands, R. (1995) *Friday Night, Saturday Night: Youth Cultural Identification in the Post-Industrial City*. Newcastle: University of Newcastle (methods section).
Maxwell, J. (1996) *Qualitative Research Design*. London: Sage.

Notes

1. 'Going out', for the purpose of the research, was defined as attending a pub/bar, club or music venue in the city centre in the evening. While night-life does not by any means subsume youth cultural activity and leisure, it is nonetheless a significant social and indeed important spatial element of one identity (see Hollands, 1995).

2. For instance, my thesis (published as Hollands, 1990), not only contained chapters on leisure and young people's use of public spaces, but also took very much of a cultural approach towards understanding labour market transitions.

3. Such a project would, I suppose, have become contemporary to numerous studies of youth as an 'underclass' conducted in the 1990s (see Murray, 1990, 1994; for critiques see Bagguley and Mann, 1992 and later MacDonald, 1997). See also the special *Youth and Policy* issue 37 (June 1992) on the riots in Newcastle.

4. While I largely reject the term 'rave culture' as a media-based, stereotypical and limited concept, used to simplify and indeed criminalize youth cultural activity, its very presence as a media-based phenomenon in the late 1980s and earlier 1990s deserved more sociological analyses than it got. For an early discussion see Redhead (1993).

5. There are many different uses of the term 'post-industrial', ranging from Daniel Bell's classic early optimistic scenario (see Bell, 1974) to more critical assessments (i.e. Kumar, 1978). My position, while leaning quite heavily on the latter approach, also gives emphases to some of the opportunities (particularly with respect to the changing position of women). For local discussions of the concept see Robinson (1988) and Garrahan and Stewart (1994).

6. For example, as MacDonald, Banks and Hollands (1993: 4) stated back in 1993: 'One of the most significant tasks facing those involved with the study of youth is to confront the "two traditions" that have crystallised in research on youth in Britain over the past twenty years . . . Whilst researchers in this second tradition have been busy surveying the structures shaping youth transitions, little has been said recently about youth culture . . . There is a crying need for researchers, in the 1990s, to turn again to more theoretically informed investigations of youth (sub) cultures which are cognisant of weaknesses and strengths in earlier traditions.'

7. This is surprising considering Willis' classic cultural study of youth transitions from school to work (see Willis, 1977). Also, I would consider my own work (Hollands, 1990) an extension of this culturalist tradition and an exception to the transition/youth culture divide. For a critique of cultural Marxism as applied to youth see Mizen's (1995) more orthodox approach. For my counter-critique see Hollands (1996).

8. For example, despite the rise of acid house/rave culture in the late 1980s, beyond a strictly conventional and empiricist chapter on youth leisure in the Banks et al. (1992) ESRC-inspired volume, *Careers and Identities* (see Jamison chapter), there was no sustained analysis of youth subcultures in any of the publications coming out of the initiative.

9. For a book which has recently tried to address this balance see Skeleton and Valentine (1998). Not surprisingly, contributors to this volume are largely geographers.

10. The Bigg Market is an area in Newcastle city centre which derived its name from a type of barley sold there and was the site of numerous busy markets, inns and hostelries from the eighteenth century. While it is still the site of a daytime market, the area was transformed in the 1980s when many of the pubs and coffee shops were bought up and altered into establishments with what can only be described as a night-club atmosphere. Today there is a high concentration of pubs there and the area has been characterized by high spirits and a somewhat (undeserved) notorious reputation for drinking, violence and promiscuity (see the discussion which follows).

11. This purely bogus criticism was made despite the obvious point that the city council had no access to such academic funds nor the expertise to bid for them. Hence the money obtained from the ESRC had no links whatsoever to cutbacks in services instituted by the council at this time.

12. In 1992 the city introduced 16 closed circuit television (CCTV) cameras in

various places in the city centre, which were linked into the police station in town. This comment concerned buying an extra camera to supplement the existing stock.

13. The term 'Geordie' while originally defined literally as miner or 'pitman', is used to refer to anyone born and bred in Newcastle and surrounding area (see Colls and Lancaster, 1992).

14. For example the *Evening Chronicle*'s coverage of the research involved a series of short interviews with young people, who were mainly critical of the research. Similarly, the Tyne Tees Television coverage included a number of young people who suggested that the study was a waste of money and of no relevance to them.

15. Hence the reference to Paul Gascoigne, whose drinking exploits and involvement in domestic violence have also graced many national papers many times.

16. The one exception to this might be the local press habit of conducting vox pop interviews with young people on an issue (which they did in this particular case: see note 14). The problem however is not only whether the young people chosen for interview are representative at all, but whether the questions they have to respond to are so highly structured and biased as to be almost wholly misleading.

17. In addition to the report and book I have also presented aspects of the research in a dozen conference presentations in various countries including the UK, Ireland, Canada, Norway and Portugal, and have published material on this topic in numerous articles in various books and journals (for example, see Hollands, 1997a, 1997b, 1999).

18. University students were much more likely than locals to report that excess alcohol consumption affected their work, created absenteeism and led to arguments amongst friends. Both student men and student women had higher weekly alcohol consumption levels (due in part to the higher number of times they went out), with student women drinking over 40 per cent more a week than local women.

19. For example, a Wellcome Trust sponsored survey of 19,000 people in the UK found that 46 per cent of men and 60 per cent of women aged 16–24 reported having only one sexual partner over the last year, while only 3.5 per cent and 1 per cent respectively had more than five (*Guardian*, 3 December 1992).

References

Bagguley, P. and Mann, K. (1992) 'Idle, thieving, bastards: scholarly representations of the underclass', *Work, Employment and Society*, 6 (1): 113–126.

Banks, M., Bates, I., Breakwell, G., Bynner J., Emler, N., Jamison, L. and Roberts, K. (1992) *Careers and Identities*. Milton Keynes: Open University Press.

Bates, I. and Riseborough, G. (eds) (1993) *Youth and Inequality*. Milton Keynes: Open University Press.

Bell, D. (1974) *The Coming of Post-Industrial Society*. London: Heinemann.

Cantelon, H. and Hollands, R. (eds) (1988) *Leisure, Sport and Working Class Cultures: Theory and History*. Toronto: Garamond.

Centre for Research on Crime, Policing and the Community (1993) *Evaluation of the Urban Crime Fund*. Report to Northumbria Police. Newcastle: University of Newcastle.

Colls, R. and Lancaster, B. (eds) (1992) *Geordies: The Roots of Regionalism*. Edinburgh: Edinburgh University Press.

Garrahan, P. and Stewart, P. (eds) (1994) *Urban Change and Renewal: The Paradox of Place*. Aldershot: Avebury.

Gayle, V. (1998) 'Structural and cultural approaches to youth: structuration theory and bridging the gap', *Youth and Policy*, 61: 59–73.

Giddens, A. (1984) *The Constitution of Society*. Cambridge: Polity Press.

Giddens, A. (1990) *The Consequences of Modernity*. Cambridge: Polity Press.

Giddens, A. (1997) *Sociology*, 3rd edition. Cambridge: Polity Press.

Goode, E. and Ben-Yehuda, N. (1994) *Moral Panics: The Social Construction of Deviance*. Oxford: Blackwell.

Gottdiener, M. (1994) *The New Urban Sociology*. London: McGraw-Hill.

Gregory, D. and Urry, J. (eds) (1985) *Social Relations and Spatial Structures*. Basingstoke: Macmillan.

Griffin, C. (1997) 'Representations of youth', in J. Roche and S. Tucker (eds) *Youth in Society*. London: Sage. pp. 17–25.

Hall, S. and Jefferson T. (eds) (1976) *Resistance through Ritual: Youth Subcultures in Post-War Britain*. London: Hutchinson.

Hall, S., Critcher, C., Jefferson, T., Clarke, J. and Roberts, B. (1978) *Policing the Crisis: Mugging, the State and Law and Order*. London: Macmillan.

Harvey, D. (1989) *The Condition of Postmodernity*. Oxford: Blackwell.

Hebdige, D. (1979) *Subculture: The Meaning of Style*. London: Methuen.

Hollands, R. (1990) *The Long Transition: Class, Culture and Youth Training*. London: Macmillan.

Hollands, R. (1995) *Friday Night, Saturday Night. Youth Cultural Identification in the Post-Industrial City*. Newcastle: University of Newcastle.

Hollands, R. (1996) 'Review of P. Mizen, *The State, Young People and Youth Training: in and against the Training State*', *Journal of Social Policy*, 25 January: 147–148.

Hollands, R. (1997a) 'From shipyards to nightclubs? Restructuring young adults' employment, household and consumption identities in the North-East of England', *Berkeley Journal of Sociology*, 41: 41–66.

Hollands, R. (1997b) 'As identidades juvenis e a cidade', in C. Fortuna (ed.) *Cidade, Cultura e Globalizacão*. Oeiras: Celta. pp. 207–230.

Hollands, R. (1997c) 'From shipyards to nightclubs: the restructuring of young "Geordie" work, home and consumption identities', in A. Mariussen, and J. Wheelock (eds) *Households, Work and Economic Change*. Boston: Kluwer Press. pp. 173–186.

Hollands, R. (1999) 'From the industrial revolution to the "fun city"?: Restructuring young adults' employment, household and consumption identities in the North-East of England', in J. Tomaney and N. Ward (eds) *A Region in Transition: North East England at the Millennium*. Aldershot: Avebury.

Hope, T. (1983) *The Prevention of Disorder Associated With Licensed Premises*. Newcastle: Northumbria Police.

Kumar, K. (1978) *Prophecy and Progress: The Sociology of Industrial and Post-Industrial Society*. Harmondsworth: Penguin.

MacDonald, R. (ed.) (1997) *Youth, 'the Underclass' and Social Exclusion*. London: Routledge.

MacDonald, R., Banks, S. and Hollands, R. (1993) 'Youth and policy in the 1990s: an editorial introduction', *Youth and Policy*, 40 (April): 1–9.

Massey, D. (1994) *Space, Place and Gender*. Cambridge: Polity Press.

Mills, C.W. (1975) *The Sociological Imagination*. Oxford: Oxford University Press.

Mizen, P. (1995) *The State, Young People and Youth Training: in and against the Training State*. London: Mansell.

Murray, C. (1990) *The Emerging British Underclass*. London: Institute for Economic Affairs.

Murray, C. (1994) *Underclass: The Crisis Deepens*. London: Institute for Economic Affairs.

Pearson, G. and Mungham, G. (eds) (1976) *Working Class Youth Cultures*. London: Routledge.

Redhead, S. (ed.) (1993) *Rave Off: Politics and Deviance in Contemporary Youth Culture*. Aldershot: Avebury.

Robinson, F. (ed.) (1988) *Post-Industrial Tyneside*. Newcastle upon Tyne: Newcastle Libraries.

Rubington, E. and Weinberg, M. (1995) *The Study of Social Problems: Seven Perspectives*. Oxford: Oxford University Press.

Saunders, P. (1986) *Social Theory and the Urban Question*. London: Unwin Hyman.

Savage, M. and Warde, A. (1993) *Urban Sociology, Capitalism and Modernity*. London: Macmillan.

Skeleton, T. and Valentine, G. (eds) (1998) *Cool Places: Geographies of Youth Culture*. London: Routledge.

Thompson, E.P (1980) *Writing by Candlelight*. London: Merlin.

Watson, S. and Gibson, K. (eds) (1995) *Postmodern Cities and Spaces*. Oxford: Blackwell.

Willis, P. (1977) *Learning to Labour: How Working Class Kids Get Working Class Jobs*. Westmead: Saxon House.

Willis, P. (1978) *Profane Culture*. London: Routledge.

Youth and Policy, 37 (June 1992).

Acknowledgements

I would like to thank Victor Jupp, who first proposed the idea of including a chapter on this topic in a volume on researching crime and for his valuable editorial comments. I would also like to thank Rob MacDonald, Vernon Gayle and Colin Clark for their comments on an earlier draft. Also a special thanks to Camilla Cowan, who provided a useful student perspective on my findings and whose humorous ramblings on 'relevant urban policy' never allowed me to take myself too seriously when writing this chapter.

12

DOING RESEARCH IN A PRISON SETTING

Carol Martin

Contents

This chapter will describe some of the practical considerations in carrying out fieldwork in prisons. It is intended to outline the most common methodological approaches and to examine what it is actually like to do research in a prison setting.

Discussing the centralization of penal institutions in 1878, Sidney and Beatrice Webb wrote 'that the prison became a "silent world", shrouded so far as the public is concerned in almost complete darkness' (1982: 278). Whilst this is not true today, the vast majority of the public still have no idea what prison life is really like and mostly draw their impressions from often misleading fictional television programmes or newspaper articles. Furthermore, many criminal justice professionals are not much better informed, although perhaps some will have visited a prison as part of their training or their job, such as solicitors on legal visits or police interviewing prisoners in gaol. Indeed, although prisons are now much more open places, and many have an almost constant stream of outside visitors coming through their doors, the true heart of prison life remains just as hidden as before. A prison has its own subculture (a set of behaviours, rules and attitudes specific to a particular group of people) which is familiar only to those who work, or live there; even language is different in a prison. Prison subculture has been the subject of numerous studies, many of them from America; some classic US literature on prisons and other total institutions include Sykes (1958), Goffmann (1968) and Irwin and Cressey (1962). In the UK, too, there have been notable studies of prison society, including those by Morris and Morris (1963), Cohen and Taylor (1972), and Jimmy Boyle, a former inmate (1977, 1984).[1]

There are various ways to become a prison researcher and most involve luck and determination. Actual routes into prison could be by doing an individual prison-based research project for a PhD or Master's degree and gaining permission from a governor to undertake fieldwork in his/her establishment. Other options may involve people who currently work or who have previously worked in prisons as, for example, teachers, probation officers or prison staff and who have decided upon a change of direction in their careers. An example of this might be a prison officer who decides to study for a degree or who wishes to become a probation officer or prison psychologist. A further possibility might be to work for a research institution which includes prison research among its interests or a government agency or department, such as the Prison Service or the Home Office. Finally, there are occasionally some experienced prison researchers who will allow a student to 'tag along' to 'learn the trade' possibly in exchange for library-based research. Prison research contracts are rarely awarded to a novice. It is something of a 'Catch 22' situation: it is very difficult to get work as a prison researcher without experience and it is very hard to access prisons to gain experience. It is a difficult world to enter before one even arrives at the prison gates. It is an unpleasant world. No television film can ever portray the real feel of a prison wing – the noise, the smell and the sense of claustrophobia. Therefore, if possible, it is definitely a good idea to visit a prison before making a decision about whether to embark on a research career involving such places.

Types of prison

The regime and conditions can vary enormously from one type of establishment to another and this may affect the type of research which can be
carried out. Prisons in England and Wales are run by the Prison Service, a
government agency, which has its own Director General and executive board of
directors. There is a prisons minister within the Home Office, but the ultimate
responsibility for all prison matters lies with the Home Secretary. Prisons in
Scotland fall within the remit of the Scottish Office and prisons in Northern
Ireland are the responsibility of the Northern Ireland Office. In spring 2000 the
prison estate comprised 135 prisons in England and Wales housing approxi
mately 63,500 men, women and young offenders. Young male prisoners
between the ages of 15 and 20 are held in young offenders' institutions and the
adult estate covers all those aged 21 and above and all female prisoners.
Juveniles under the age of 15 cannot be held in prison but may be sentenced to a
period of custody in a secure training centre, for 12–14-year olds; male young
offenders and adults may not be held together. New rules governing juveniles
and young offenders will come into operation during the year 2000.

The vast majority of prisoners are young men who are both socially and
economically disadvantaged and whilst the female prison population has been
rising over the last few years, women still represent only a small fraction of all
those incarcerated (4.25%). Prisons can be broadly divided into two main
types: local and remand institutions and training prisons. The former deal
with prisoners awaiting trial or sentence (on remand), those serving short
sentences and those who are being assessed before being sent elsewhere. The
large Victorian prisons in city centres such as Pentonville, Wormwood Scrubs,
Wandsworth, Strangeways, Liverpool and Winson Green (among others) are
all local prisons. Apart from holding prisoners on remand or those serving
short sentences from the local area (hence the name 'local' prison), they also
hold men starting long sentences on 'induction units' who need a period of
stabilizing and careful watching before being moved on through the system.
Training prisons can be 'open' or 'closed' and are categorized according to the
security level they provide from Category A establishments which are the
most secure, to Category D which means 'open' conditions. It is actually the
prisoners themselves who attract a particular security category: this is based
on an assessment of their level of dangerousness to the public, or the state, or
the likelihood of their escape, a system which was introduced following the
Mountbatten Report in 1966. Thus a multiple murderer or high profile spy
who had committed crimes against the state would attract a 'Cat. A' status
whereas someone convicted of a white-collar crime such as tax evasion or
fraud would probably be categorized as 'Cat. D'. An 'open' prison literally
means there are no physical barriers between the prison and the outside world
– no walls, gates or reinforced perimeter fences. Prisoners serving sentences in
such establishments are trusted not to abscond and would probably carry out
voluntary work in the local community.

Once within the prison system, prisoners serving sentences of several years
have a sentence plan, which is drawn up by the prisoner and his/her personal

officer – the prison officer who has special responsibility for that prisoner within the establishment. A sentence plan should include realistic, achievable aims for a prisoner in terms of courses they should try to attend, such as in anger management, drugs or social skills; it also details whether a prisoner has particular educational needs and where s/he would ideally prefer to be located to facilitate visits from his/her family. During the course of a sentence, especially if it is of several years' duration, many prisoners move from a prison operating a fairly restrictive regime to a more liberal prison as their sentence progresses. Apart from these 'progressive' moves, prisoners can also be transferred between prisons because of overcrowding or for disciplinary reasons. Prisoners who are disruptive, difficult to manage or who have a bad disciplinary record are frequently moved between prisons in a practice which is variously known as 'the ghost train', 'magic roundabout' or 'the shared misery circuit'. This practice is highly unpopular among prisoners and was heavily criticized in the Woolf Report (1991), and whilst it is probably less common than it used to be, transfers between prisons solely for disciplinary reasons still continue.

When planning what methodological approach to use for a research project, the type and security level of the proposed prison are important. An open prison offers the greatest freedom of movement for researchers with the minimal involvement of prison officers, whereas in a high security establishment, such as a dispersal prison, which is one to which a 'Cat. A' prisoner may be transferred, an escort may be needed. This means that the governor of a dispersal could be less likely to want a research project within its walls if there are resource implications in terms of staff time for escort duty. Furthermore, the high security classification of individual prisoners may mean that, for safety reasons, staff could be unwilling to allow inmates to be interviewed without a staff member present, which would impact on the confidentiality of the research.

Planning research

When a prison research project is in the planning stages it is vital to remember the most obvious fact of all – individuals are not allowed inside a prison simply because they want to get inside, for whatever reason. Permission must be sought from either the Prison Service or the governor of an individual establishment. Therefore it is important to decide what type of research project is to be carried out and, crucially, whether it is likely to meet with official approval.

A number of different types of studies are carried out inside prisons. The Prison Service has its own research and statistics department and carries out much of its own research. Examples of this in recent years include: a study on imprisoned women and mothers; an examination of the effectiveness of day visits to prisons for young offenders, sometimes called 'scared straight programmes'; issues of control in Category C prisons; parenthood training for young offenders; the welfare needs of unconvicted prisoners; and a study of reconviction rates following release from a therapeutic prison. The Service also commissions a number of studies from universities and independent research agencies where perceived independence in the research is an important factor. Recent examples of this type of study include: an evaluation of a drug rehabilitation course operating in a prison; an evaluation of a private remand establishment; victimization in prison; the effectiveness of mandatory drug testing in prisons; suicide and self-harm; an evaluation of prison incentive schemes; and an examination of prisoners who are difficult to manage.[2] The Prison Service may also grant permission for recognized external bodies, such as university departments or some charities, to carry out research on topics of their own choosing. Such studies include research on fights and assaults between inmates, drug rehabilitation programmes and staff/prisoner relationships.

If permission has been granted by Prison Service headquarters to carry out research, then the governor's approval is usually a formality. Governors may also grant access to 'their' own establishment if the proposed research is likely to cost them nothing and could be of use to them by addressing a current issue. This last route is probably the easiest for a small scale project because it bypasses the bureaucracy of Prison Service headquarters. Apart from size, the crucial factors determining the viability of a proposed project are: finance, value, feasibility or degree of difficulty, ethics, local disruption and politics and hidden agendas. All of these factors need to be considered when drawing up a prison research proposal (see Francis in Chapter 2 for a more general discussion of factors which influence research planning).

Finance

The recipient of funding, say from a university or a grant making body, does not need to approach the Prison Service for financial support of any sort and consequently the battle is already half won. At the time of writing, contracts awarded by the Prison Service are subject to competitive tendering if the value is above £10,000. Unsolicited research proposals submitted by outside researchers are sometimes approved and financed by the Service but the scope of the research must fall within the overall research agenda of the Prison Service. This is not a secret agenda, and ideas for research can be informally discussed with the Service to find out whether there may be an opportunity to proceed to the next stage. However, even if an original idea fits within the overall Service research programme, an innovative proposal could be submitted to the Service only for the originator to find that it would need to be put out to tender and it could then be awarded to someone else altogether. (This, of course, does not just apply to the Prison Service but is a difficulty in all social science research sponsored by government departments.)

Value

In this context, 'value' means *outcome value* to the Prison Service, not value for money. If the proposed research findings would represent no intrinsic value to the Service, the research is unlikely to be sanctioned, no matter what value for money it might represent. The research question must be one that will have some kind of policy or practice impact for the Service or the chosen establishment. For example, a governor may know that in his/her establishment victimization (or bullying) is a particular problem. Perhaps there have been a number of incidents of self-harm which staff are convinced could be attributed to assaultive behaviour by some inmates on others who are more vulnerable. The governor may have money in the prison's budget to commission a study him/herself or, more likely, if an outside research body wanted to carry out a study on that subject, the governor would probably be delighted and the research team would be welcomed with open arms. To ensure that the research represented value for the establishment, the governor might suggest specific research questions s/he would like examined or suggest new policy approaches, the adoption of which would hinge on the research findings. There are many ways in which questions which are of interest to criminologists could have practical and policy implications for the everyday running of establishments. When planning research, a key factor is being able to identify the ones which are most likely to represent the best value available to the prospective client, that is, the Prison Service as a whole or an individual establishment.

Feasibility or degree of difficulty

The research must be 'do-able'. However interesting or important the findings might be in terms of policy or practice, if the research is viewed as impossible or very difficult to carry out, it is unlikely that permission will be granted. An example of such projects might be the impact on mental health and personal safety of sexual assaults between inmates in prisons, sexual relationships between inmates and staff, or the level and seriousness of unreported assaults by staff on prisoners. Such studies would have serious implications for healthcare provision, mental health, discipline, and suicide and self-harm, but would be extremely difficult, if not impossible to carry out. Apart from issues surrounding the highly sensitive nature of such disclosure by prisoners, the Prison Officers Association has to date been very resistant to research that might uncover undesirable conduct by its members. A governor is unlikely to sanction research if there is the prospect of opposition from within staff ranks. Few, if any, large scale projects of this type have ever been conducted in this country.

Ethics

The research must not harm or put the subjects at risk in any way. There is an ethics committee at the Prison Service which examines all research proposals that might give rise to ethical concerns. For example, any research which asks prisoners or staff to disclose illegal behaviour (such as drug use or assaultive behaviour) would be carefully scrutinized. (Other ethical issues concerning prison research will be discussed in more detail later in the chapter.)

Local disruption

If conducting research within an establishment would involve significant disruption to the regime or the daily running of the prison, it is most unlikely to be sanctioned. Most research has to fit in with whatever else is going on. It must be remembered that any prison research will have some kind of disruptive effect on someone within the establishment, however benign. The impact on staff time is also a major consideration. The greater the disruption, the more important and high profile the research needs to be to gain approval.

Politics and hidden agendas

Finally, there is the influence of politics, which is the most difficult to assess. It is difficult for two reasons: first, the current political agenda in broad terms, that is nationally, is important; and secondly, the 'hidden agenda' within the Prison Service *at any given time* is crucial. Some research might never be approved because it is considered too theoretical and is not recognized as having an application in practical terms, or perhaps the underlying theoretical model may not be politically acceptable at a given time. For example if a research project had been proposed several years ago to examine Michael Howard's assertion as Home Secretary that 'prison works', it would probably not have been officially sanctioned and the Home Office would certainly not have funded it. The problem is that unless one is an 'insider', one does not know what is currently acceptable. Such 'insiders' include Prison Service staff or Home Office employees and those from a variety of disciplines who are in positions of influence and are consulted over criminal justice policy. These include representatives from probation, police, the courts, the legal profession, senior academics and pressure groups like the Penal Affairs Consortium, Prison Reform Trust, and the Howard League.

Gaining access

This section could, perhaps, be subtitled 'Negotiating gatekeepers'. Some considerations in obtaining permission to work within a prison have already been covered. The prospective researcher may think that once inside an establishment, access problems have been left behind at the prison gates; unfortunately this is not usually the case. This section looks at some of the practical obstacles to gaining access to a prison establishment and its inmates.

Once permission has been granted to allow access to the establishment and to the subjects, a researcher's status will determine, to some degree, the *level* of access within that prison. In general terms, experienced prison researchers who have a track record of working for the Prison Service are likely to be accorded a high degree of co-operation at management level within the establishment. Thus a study commissioned by the Home Office or Prison Service and employing Home Office researchers would probably gain the highest level of official co-operation, enabling carrying of keys, and facilitating access to all necessary parts of the prison and prisoner records. If problems are being encountered by the researchers, such as officers refusing to unlock subjects for interview or denying access to records or staff lists, help can be enlisted from the governor of the establishment and his or her authority invoked to support the researchers. Nevertheless, this approach in dealing with such complaints would not necessarily promote a happy or constructive atmosphere between the researchers and the uniformed staff and would probably be counter-productive. A high level of management co-operation does not always filter down to the wings, and staff and prisoners alike may be suspicious or even hostile at the outset of a research project. The individual skills and experience of the researcher in negotiating acceptance of his/her presence and of the nature of the research can determine the success or failure of the whole project. If the research is oriented towards prisoner 'welfare' issues, staff may be resentful and reluctant to facilitate access to prisoners or office space for interviews; if the research requires lengthy periods of time to be spent talking to and getting to know staff, then prisoners may decide the researchers are 'on the staff's side' and thus not to be trusted.

Once the researcher is finally face to face with someone from the subject group, great care should be taken to explain the research fully and to stress the voluntary nature of participation. Obtaining *informed consent* means being sure that anyone who is a prospective subject understands what they are participating in and what implications this could have for them. All prison research which entails interviewing staff or prisoners must be voluntary on their part; coercion should never be used to engage participants in a study and prison officers should not be asked to explain the research to inmates or ask them to agree to an interview. It should always be made clear at the outset of any interview that participation carries no reward or disadvantage. Despite public perception to the contrary, prisoners are an extremely vulnerable group of people, without the freedom to walk away from a situation or the normal recourse to legal help and advice. They cannot pick up the telephone at will or pop into a friend's house to discuss what they should do. They may feel that to co-operate might be detrimental to their position but that not to co-operate could have even worse consequences for them. They are a powerless group of people about whose lives others take all the decisions, whether they like those decisions or not.

Getting into and around an establishment

Before actually starting work in a prison, all outsiders are given a security talk and most prisons will insist on giving such a talk to anyone new to that particular prison even if they have worked in other prisons before. These talks cover such issues as information/intelligence, prohibited items, carrying keys, personal safety, alarm bells, dress code, carrying contraband, taking care of belongings, what to do in an emergency and confidentiality or security of information. Security officers usually impress upon all newcomers that they are an 'extra pair of eyes and ears' and that anything suspicious must be reported to them, perhaps to add that 'missing piece' to the intelligence jigsaw of prison life. On hearing this security talk, a new researcher is faced with the possibility of the first conflict of interest s/he may have to deal with during the course of his/her time in the prison   confidentiality to the prisoner or security information for the staff. This is not an easy issue to resolve and the researcher will be constantly aware of this underlying tension between con-fidentiality and security during the entire fieldwork period in a prison. It can be helpful at this stage, if appropriate, for the researcher tactfully to outline the terms of confidentiality under which s/he is operating.

Security staff will also warn against bringing anything in or taking anything out of the establishment for a prisoner. This does not just refer to prohibited or restricted articles, such as tobacco, drugs and alcohol, but could mean anything as seemingly innocuous as posting a letter. In certain circumstances, researchers can be issued with a set of keys to facilitate freedom of movement in a prison. This is particularly appropriate where the researcher needs to move between prison wings or different buildings during the course of a day. Most prisons do not have spare members of staff available to escort researchers around an establishment and many hours can be wasted just by waiting to be let through locked gates and doors. The most significant disadvantage of carrying keys is that researchers may be identified by the prisoners as being part of the establishment, thus compromising perceived independence. Cell keys are never issued to non-prison staff, thus minimizing the risk that they will be taken hostage, or that other potentially difficult security situations will arise. The major practical disadvantage of carrying keys is that if any mistakes occur they are likely to be of serious proportions, and even the most trained and experienced staff members make mistakes. Changing the locks in a prison can cost many thousands of pounds, but it must be done if there is a breach of security. Apocryphal tales about mistakes with keys abound and all staff have heard of prisoners who can memorize a key pattern from a mere glance of only a couple of seconds' duration. More than one prison in recent years has had to change all the locks in the entire establishment when keys have been shown inadvertently during the broadcasting of television documentary films or information videos for new prisoners.

Confidentiality for the prison is a different issue entirely from the questions of participant confidentiality in research. Prison security staff must be satisfied that outsiders understand that they may not breach the confidentiality or security of the prison or disclose any information about its staff or inmates to others. Whether or not an individual has actually signed the Official Secrets Act, anyone working in a prison is bound by its provisions and so prospective researchers should familiarize themselves with the contents of the Act and how it would apply to them.

Safety

People who are in prison do not, on the whole, want to be there and some of them have committed extremely serious and violent acts. In this country, prison is the most severe method of punishment available to the courts. It has been said that prisons house the mad, the bad and the sad, although only the bad should be there. Certainly some inmates are extremely dangerous and continue to commit sometimes serious crimes while they are in prison. Theoretically, the higher the security classification, the more dangerous the prisoner. But most prisoners, most of the time, just want to get on with their sentences and get out. Many Category C or D establishments will allow inmates access to craft materials and tools such as Stanley knives, scissors or DIY implements, certainly in a working environment and sometimes in their cells. Other prisons will not even allow metal eating objects, and for particularly difficult or violent prisoners, in-cell tables and chairs are made out of reinforced cardboard. Most prisons display a collection of weapons that have been made out of the most ordinary everyday objects. Some weapons commonly used in prison include a sock containing a PP9 battery or a glass jar full of jam or margarine, which is used as a cosh; bladed weapons fashioned out of toothbrush handles with razor blades embedded in the end; syringes secreted inside pens; and boiling, sugared water.

Researchers should be safety conscious at all times in prisons, but especially when interviewing in a one-to-one situation. Generally, prisoners should be interviewed in an office where an alarm bell is within easy reach, and always with the knowledge and permission of the landing officers. A visitor to any part of the prison must inform the senior officer in charge of his/her arrival and departure. In Category D prisons, interviews would probably take place in the prisoner's room, but again, always with the knowledge and consent of the landing or wing staff.

A researcher should never reveal personal information about him/herself, such as address, where family members go to school or where the researcher spends his/her leisure time. Off-duty officers have, occasionally, been assaulted by friends or relatives of inmates because plans for a night out were overheard. Generally, however, assaults on people who work in prisons (with the exception of uniformed officers) are extremely rare – and it is very unusual for researchers to feel unsafe or threatened in any way. In fact, prisons are not the humourless places one might expect: there is a style of grim humour which is part of prison life and it is legitimate for everyone who works or lives there to be part of it. Everybody has their bad days; staff and prisoners are no exception, but most prisons operate with a sort of siege mentality and working in them is by no means a dispiriting experience.

Different methodological approaches

Most social science research methodologies can be used in a prison setting but the most frequently used forms are probably observation, documentary research, self-completion questionnaires and interviewing. The decision about research methods will largely depend on the type of questions the research sets out to ask. Is the research largely collecting qualitative or quantitative data? Is demographic information needed and is external validation necessary? How best can these approaches complement each other?[3]

Observational research could probably be best described as the 'hanging out' school of research. It is not only desirable, but necessary to spend some time in an establishment before embarking on interviews with inmates or staff whether or not this forms part of the data collection, and it is important to get to know the geography of the place. Many prisons are large and complex institutions and it is very easy to get lost in labyrinthine corridors or among similar-looking outbuildings. It is easy to ask others the way, but it does nothing for the researcher's feelings of professional competence and self-confidence, notwithstanding issues of safety, to keep getting lost. Furthermore, getting to know the establishment's routine and regime is essential – what days different wings do their laundry or have access to the prison shop (canteen)? What time of day do inmates go to the gym, have visits, association (free time) or exercise? Are there any half-days when all prisoners are locked up and the prison is at patrol state (i.e. where only one or two officers are on duty and prisoners are unlocked only in the most extreme emergency)? Once the researcher is a familiar face around the prison, and talked to by both prisoners and staff, it is much easier to ask for interviews. It is also vital at an early point to stress the confidentiality and independence of the research, if that is appropriate. It is wise to be as honest and open as possible about the research; over long periods in a prison, it is surprising how much other people remember. Researchers may be trained to remember detailed information for their study, but they often forget they are under as much scrutiny as their subjects. Prisons are like goldfish bowls – everything that happens is seen and talked about by a large number of other people.

Triangulation

Triangulation of method – that is, the use of different methods to study the same phenomena – can be immensely valuable. For example, observational techniques combined with interviewing and documentary research allows the researcher to start with some preliminary observation, move into the interview phase whilst conducting the documentary research alongside, and conclude with more observation. This is illustrated in Figure 12.1.

In this way, issues which arise during the course of the fieldwork can be clarified before the end of the research period. If all the observation is done at the outset of the fieldwork, followed by a period of two or three months' interviewing, it can be very difficult to remember even everyday routines once the researcher has left the prison. Field notes are vital: they are an informal diary of events and personal impressions which should be kept up to date and can act as a powerful memory aid. They also form part of reflexivity whereby the researcher can reflect upon the research process and his or her effect upon it and its outcomes. There is an unstated protocol about writing field notes: this should be done in privacy, possibly at the end of the day and certainly out of sight of the research subjects. If observing something as sensitive as group therapy, researchers should never write down anything, or be seen writing up notes afterwards.

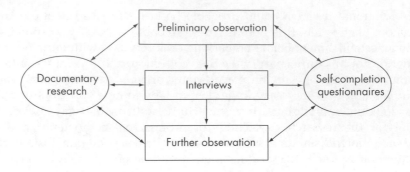

Figure 12.1   *Triangulation of methods in prison research*

Interviews

Interviewing can be formal or informal, following structured or semi-structured questionnaires or taped for transcription and analysis afterwards. Structured questionnaires take the longest preparation time, but are probably the easiest to analyse afterwards; they are not as rigid as their title implies and open and qualitative questions are often included. Taped interviews sound the easiest to conduct, which they undoubtedly are, using a prompt or crib sheet to ensure the questions are asked in a consistent manner. They allow a flow of conversation that would be inhibited by a researcher laboriously writing everything down. This type of interview is probably best for life history work. However, the inescapable disadvantage of taped interviews is the hugely time-consuming transcription and analysis. Self-completion questionnaires can be useful if a large scale survey of a prison is intended, and some prisoners feel there is the added insurance of anonymity. This method can often successfully be combined with follow up, in-depth interviews. As with all methodologies, the type of research question will largely determine which form of interview to use. Whatever type of questionnaire is planned, it is always advisable to conduct a pilot study (a trial run, using the planned survey method with a small sample). This allows adjustments or changes to be made to the interview schedules without using the target group.

Case selection

Who to interview should be determined in advance. A *random sample* means selecting participants, either alphabetically or numerically, to represent a population, taking no other factors into consideration. In a prison, if random sampling is chosen, the prison roll can be used (a computerized daily printout of all inmates in the jail in alphabetical order) to select every third or fifth (or whatever) prisoner. This provides a good starting point. A helpful officer with access to the LIDS (Local Inmate Data System) computer will prove an invaluable asset. This computer system contains information on all inmates in an establishment with details of where the prisoner is housed, whether s/he works and where, date of birth, parole details, release date and so on. *Opportunistic sampling* is a method by which subjects are approached *if they happen to be there*. This does not give a random sample, however, and may have disadvantages. For example, prisoners who are always locked up on their wings may be unemployed and have chosen not to work, or may be subject to disciplinary sanctions, which might skew the sample towards a certain type of prisoner. Opportunistic sampling is probably more representative with a staff sample because shift patterns, not choice, determine whether staff are on duty at a particular time. *Snowball sampling* is recruiting interview subjects by word of mouth, enlisting prisoners' help by asking those who have already completed an interview to make an introduction or recommend a friend to take part in the research. This can be a useful approach in prisons, as someone who has been recruited on the interviewer's behalf by a friend may be more likely to participate in the research than a prisoner who has been approached 'cold'. Like opportunistic sampling, this is not a truly representative method of recruitment, but there can be significant advantages. Depending on the topic being studied in the research, no one is in a better position to know what is going on in a prison than the prisoners themselves. If an interviewee relates anecdotal information concerning another inmate, they are perfectly placed to approach that inmate for an interview, whereas the researcher could not reveal that s/he had information about them. Furthermore, if a particular population is being studied, for example drug users, snowball sampling would offer the best opportunity to access those within a specific group. With snowball sampling, it is crucial to get the initial approach right and the prison grapevine should never be underestimated. The first few interviews can virtually determine the success or failure of the interview phase; a couple of successful interviews with well-chosen subjects at the outset can be an invaluable recommendation to others to take part.

In terms of credibility and trust, an independent researcher from a recognized external body such as a university or charity has a head start in gaining the confidence of the prospective subjects. This may not be required for all types of research, and if sensitive questions are not being asked, it is not necessary to stress any kind of confidentiality.

Documents

Documentary research using prison records can validate information gained through interview. A prisoner's individual prison record will hold personal details such as date of birth, sentence length, previous convictions, disciplinary record and custodial history, some or all of which is likely to be of use to the researcher. As with most self-reported information, the details prisoners reveal about themselves tend to be accurate in the vast majority of cases. It is unusual for prisoners to supply misleading information about themselves or their offending history; this does sometimes occur if a prisoner has committed a sexual offence which s/he does not wish to reveal. When accessing documentary records, it is not unusual to find key pieces of information missing from files or whole records unavailable for scrutiny by researchers; a number of different people in prisons use inmate records and sometimes these simply cannot be accessed when the researcher needs them. It is helpful to have written permission from the governor to access inmate records to be able to show administrative staff upon demand.

Ethical questions

Ethical questions that need to be considered when undertaking prison research include the vulnerability of the subject group, issues of gender, the level and extent of confidentiality and the researcher's status. One distinct advantage of carrying out research among such a literally captive audience is that the refusal rate is usually low. Prison is an excruciatingly boring place and the boredom often makes prisoners willing to participate. Once the research has been explained to the inmates and staff, and if the prisoner understands that s/he *need* not comply with the request for an interview, then consent to participate is very much dependent on the individual skills that the researcher needs to employ. Can the researcher offer *anything* the prisoner wants? It is very unlikely that the inmate will agree to an interview if the answer is no. Prisoners agree to be interviewed because they will get something out of it for themselves. This is known as the *research bargain*. Whilst no inducements may be offered, that does not mean the prisoner does not weigh up whether there is likely to be any other advantage attached to the interview. For the researcher, working out exactly what that advantage might be can be complex, but it is the key to why inmates agree to be interviewed and then often reveal deeply private information. In some circumstances, the bargain may be as straightforward as the prisoner simply wanting to talk to someone different about almost anything other than their immediate surroundings.

Also within the area of vulnerability of the subject group would fall the problem of identification by staff of prisoners taking part in the research. An example of this might be research concerning drug users in prison. Perhaps the most difficult thing about engaging prisoners in drugs research is that the researcher, simply by talking to them, identifies them as involved in some way. A negative consequence could be that the prisoners are then targeted for a drug test. If they are indeed using drugs, they may assume that the test was a consequence of their co-operation in the research and the word would get round very quickly for others not to take part. But if such prisoners do consent to be interviewed, some will certainly admit to drug use both inside and outside prison. A problem along similar lines is that both staff and prisoners sometimes ask researchers directly what other people have said or why certain individuals have been interviewed. In such cases, the researcher can politely remind the questioner that the same level of confidentiality is accorded to everyone.

Gender

The issue of gender in prison research is a difficult subject to confront. A feminist criminological critique might suggest that even conducting such a debate unnecessarily sexualizes the issue and marginalizes the purpose of the research and the skills of the social scientists. It might be argued that male prisoners show more interest initially in talking to female researchers, but it is highly unlikely that there is any difference in the recruitment rates for interviewees between male and female researchers in either male or female jails. It is the skill, not the gender, of the researcher that establishes his/her credibility and this is the crucial factor in determining the willingness of prisoners to participate in a research project.

Confidentiality

Researchers will both hear and be told things which may disturb and distress them, and everyone who works with prisoners and staff will probably at some time question their own need to maintain total confidentiality. Such information may include details about previously undisclosed offences or actions that have occurred while the inmate was in prison.[4] This is a grey area where there are no rule books to consult about whether a researcher should break confidentiality, but a research protocol (a set of guidelines concerning such matters, drawn up between the researchers), agreed before the commencement of the study, can help determine where such a line should be drawn. At the start of any interview, the researcher should make it quite clear to the subject what the rules of confidentiality are, so the prisoner or staff member understands exactly what s/he can safely reveal without the danger that unwelcome action will be taken against them. A decision to breach confidentiality is not a decision which should be taken lightly – an entire research project, months of work and thousands of pounds could depend on a hasty or unwise course of action. On the other hand, if information impinging upon the security of the prison was disclosed to a researcher which was not acted upon, very serious consequences might ensue. Information can be passed to the researcher which is deliberately misleading, by either staff or prisoners, in order to get someone else into trouble or to test the confidentiality of the research.

Researcher's status

The researcher's status and independence may also be called into question if a prisoner or staff member asks for help or advice. It is always difficult for a researcher to maintain a neutral position when asked to intervene in a specific situation, particularly if the researcher feels strongly it is only 'right' or 'fair' to do so. This is probably one of the commonest and yet most difficult ethical dilemmas the researcher will need to deal with. Examples of these types of situation could include the following:

- a researcher being warned that a prisoner is at risk of attack;
- a researcher having information which would materially affect the outcome of a disciplinary adjudication;
- a member of staff openly discussing having assaulted an inmate;
- a prisoner admitting s/he carried out an assault for which s/he was not charged;
- a prisoner or staff member revealing they are seriously depressed or in danger of self-harming;
- a prisoner asking a researcher to bring into or take out of the prison any items, whether contraband or not;
- a prisoner asking for help or advice with practical problems such as housing, employment or drugs counselling.

This is not an exhaustive list and is only intended to give a flavour of the sorts of dilemmas that arise for researchers.

Another important ethical issue which can impact on prison research concerns personal involvement and relationships with staff and prisoners and this will be discussed in the next section.

Problems that can be encountered

Among the more routine, yet nonetheless consequential problems which can occur in a prison include prospective interviewees not being available. Prisoners may have visitors, be at the gym or have been sent to work, despite a request for them to remain on the wing; they may be transferred to another jail with no prior notice. In a remand prison, inmates may not return to the establishment following a bail hearing in court. Staff might not be on duty. Flexibility is essential and contingency plans should always be in place with alternative work or interviews lined up. This is such a frequent occurrence in prison that arranging an alternative daily timetable becomes automatic. If travelling to a particular prison to interview only one or two people, it is advisable to make sure in advance they will be there, and available, when the researcher arrives.

No observational research can ever be value free, however hard one tries to be objective. But a trained social scientist should be aware of this at the outset and strive to eliminate personal values and assumptions from the research. In all social science research, the researcher is simply a recorder of information provided by others. Researchers are not creating the circumstances they are examining and they have a duty to impact as little as possible on existing conditions. By simply being there, researchers become part of the interaction they are observing and can affect the social situations they are studying. If this extends to the subjects actually changing their behaviour *because they are aware they are being studied* it is known as the Hawthorne effect, after an experiment carried out in America in the 1930s.[5] *Going native* is an extreme version of the researcher losing his/her objectivity and neutral perspective. The expression means personally identifying with some or all of the research subjects being studied. This type of problem is more likely to occur when researchers spend long periods of time with an isolated population such as studying long-term prisoners over a period of years, rather than weeks or months.

Another problem that occurs from time to time is personal involvement with individual prisoners or staff. In any job, it is impossible to work for any length of time with the same people and not build personal relationships with them. Prisons are not fundamentally any different. As in any other working environment, researchers will meet people they like and people they do not like among both staff and inmates but in the highly charged atmosphere of a prison they must be constantly vigilant to maintain the strictest professional conduct at all times. Any relationship which develops in a prison is, by its nature, unequal. Prison is not the real world, and in terms of personal choice, prisoners are powerless. The slightest indication of particular friendship or favouritism may cause serious problems. Nothing could be more humiliating for a researcher than to be questioned by security staff or a governor because of rumours of some kind of impropriety.

Conclusion

Prison researchers do not constitute a large body of people in the UK and there are few published articles which talk in depth about what working in a prison might have in store for the researcher. Most studies carried out in prison include a section on methodology, but this is usually confined to the technical details of the samples. This may be due to a tendency to guard jealously the 'tricks of the trade' which only experience teaches. It is probably far more likely that the potential readership of the research findings has little interest in the practical details of how the research was carried out. However, it is vital that anyone contemplating prison research gets to know such 'tricks of the trade'.

When considering whether to undertake prison research, the prospective researcher should remember that s/he will meet some people who have committed extremely unpleasant crimes. If the prospect of such encounters seems problematic, perhaps a prison is not a suitable place for that person to work in. Inmates, particularly those serving long sentences, become masters of observation and they will see through any mask put on to hide fear or revulsion. By choosing to enter the prison gates the researcher must leave his/her personal prejudices behind. Almost all prisoners will one day be released and every contact they have with the outside world during their sentence should be a positive experience. The extremely sensitive nature of the work requires careful preparation and a sound understanding of what is involved. Badly prepared and ill-informed researchers can cause a great deal of damage and the possibility of real harm should not be underestimated.

Suggested readings

Boyle, J. (1977) *A Sense of Freedom*. London: Pan.
Cavadino, M. and Dignan, J. (1992) *The Penal System*. London: Sage.
Jupp, V.R. (1989) *Methods of Criminological Research*. London: Allen and Unwin;
  reprinted by Routledge, 1996.
O'Connell Davidson, J. and Layder, D. (1994) *Methods, Sex and Madness*. London:
  Routledge.
Stern, V. (1989) *Bricks of Shame*. London: Penguin.
Sykes, G. (1958) *The Society of Captives*. Princeton: Princeton University Press.

Notes

1. For a description of the major issues in the ongoing debate about prisons and a comprehensive bibliography on prison literature, see Morgan (1997).

2. All these studies, both 'in house' and those commissioned 'externally', are available from the Home Office Research Development and Statistics Directorate publications department.

3. A detailed description of different methodologies in prison research, but particularly *participant observation*, can be found in Genders and Player (1995).

4. In general terms, there is no statutory obligation on researchers to reveal previously undisclosed criminal actions to the police unless required by them to do so. The legislation covering this is the Criminal Law Act 1967 s.4 and s.5 although this is not the same in Northern Ireland, where different statutes cover terrorist offences.

5. A full account of the experiment can be found in Roethlisberger and Dickinson (1939) *Management and the Worker: an Account of a Research Programme Conducted by the Western Electric Co., Hawthorne Works, Chicago*. Cambridge, MA: Harvard University Press.

References

Boyle, J. (1977) *A Sense of Freedom*. London: Pan.
Boyle, J. (1984) *The Pain of Confinement*. Edinburgh: Canongate.
Cohen, S. and Taylor, L. (1972) *Psychological Survival: The Experience of Long-Term Imprisonment*. London: Penguin.
Genders, E. and Player, E. (1995) *Grendon: A Study of a Therapeutic Prison*. Oxford: Clarendon Press.
Goffmann, E. (1968) *Asylums*. Harmondsworth: Penguin.
Irwin, J. and Cressey, D. (1962) 'Thieves, convicts and the inmate culture', *Social Problems*, 10 (92): 145–155.
Morgan, R. (1997) 'Imprisonment: current concerns', in M. Maguire, R. Morgan and R. Reiner (eds) *The Oxford Handbook of Criminology*, 2nd edition. Oxford: Clarendon Press.
Morris, T. and Morris, P. (1963) *Pentonville: A Sociological Study of an English Prison*. London: Routledge.

Sykes, G. (1958) *The Society of Captives*. Princeton. Princeton University Press.
Webb, B. and Webb, S. (1982) *English Prisons under Local Government* (1963) quoted in M. Fitzgerald and J. Sim, *British Prisons*. Oxford: Basil Blackwell.
Woolf Report (1991) *Prison Disturbances April 1990: Report of an Inquiry*. London: HMSO.

13

UNDERSTANDING THE POLITICS OF CRIMINOLOGICAL RESEARCH

Gordon Hughes

Contents

When one reads criminological research in most books it may seem that the research process follows a very systematic and detached logic informed by the particular intellectual and methodological premises of its line of inquiry. There is, of course, much truth in this broad impression which we gain from studying the 'finished product' of the research. However, this rather antiseptic picture may, by itself, give a dangerously skewed and false impression of the realities of actually 'doing' criminological research. All too often research publications fail to tell us about the hidden difficulties, constraints and limitations behind the apparently smooth and detached appearance of the research process. To redress this balance the aim of this chapter is to provide an informed and critical understanding of the varying political contexts and ethical dilemmas which have an impact on contemporary criminological research.

Researching in a political world

All social science has a political dimension, in the non-party-political sense. All aspects of research necessarily involve the researcher in both the analysis and practice of power and, in turn, have the potential to generate conflicts of interest between a whole host of interested parties. It will become clear that the social scientific study of people and institutions will inevitably and necessarily intrude on privacy at times, may subvert deeply held moral beliefs and practices, and may uncover unwelcome truths for some groups that are very welcome truths for others. Accordingly, no criminological research takes place in a political and normative vacuum.

There is a second, more restricted meaning of the word 'political', relating to explicit political ideologies and organized coercive institutional power of the modern nation-state. This second sense of the word 'political' is of crucial importance for research agendas and projects in criminology, given that so much criminological investigation is in fact the study of state power in its most overt form, for example the imprisonment of offenders. Indeed much of the work of the critical criminologist both challenges and subverts dominant and commonsensical assumptions about the world. Nonetheless, criminological knowledge may also be used to lend 'scientific' credibility to prevailing dominant ideological assumptions and institutional practices, for example, class, 'race' and gender inequality (see for example both the 'old' eugenics and 'new' genetics research on locating the pathological characteristic(s) marking off criminals from the 'normal' population: Hughes, 1998: Chapter 3).

Criminological research does not take place in a political and moral vacuum but is a deeply *political* process. This deeply political process paradoxically often remains hidden in the published product of research. At best, there may be a chapter or an appendix on the awkward realities of research. However, more often it is in the brief preface to the book that the reader gets a hint of the sponsor and the important contacts which underpin the research project. In the following sections the key stages of the research process are examined, drawing on illustrative examples from influential research projects. For the sake of analytical clarity, the following six stages may be distinguished (in reality it is not possible to separate out these stages so neatly from each other):

- getting started
- gaining support and sponsorship
- gaining access
- collecting the data
- publishing the results
- utilization of the research.

Getting started

All research involves the asking of questions about a particular problem and even at this initial point of departure for criminological inquiry it is impossible to avoid politics. There are instances when we may not be allowed to ask certain questions, as in a situation defined as involving 'national security'. Thus, the role of the British Forces' Special Air Service (SAS) in Northern Ireland, and its possible links to criminal assassinations of suspect 'terrorists' (the so-called 'shoot to kill' policy in the 1970s and 1980s), is an important criminological research proposal which would have been extremely difficult and dangerous to try to investigate. There is little doubt that the state would have explicitly vetoed any such proposal. Of course, the influence of state power is not always so transparent. Power is often most effective when decisions are not explicit but when influence over others is hidden. Political influence may be most telling when no overt decision to say 'no' has been made. In the context of criminological research, the influence of dominant ideologies and institutional practices may be most evident when researchers do not even think to ask 'awkward' questions.

The very formulation of a research question or hypothesis is surrounded by the political constraints noted above as well as the obverse of this, namely political opportunities. Political currents and counter-currents in any given socio-historical context do not just wash over a research culture but instead help construct its agenda. Feminism is an example of a political current which has drastically reshaped the research agenda in recent criminology (see Walklate, 1998: Chapter 5). Where not so long ago criminology was largely 'gender-blind', it is now widely acknowledged that gender differences in criminalization and victimization are crucial to criminological analysis.

Yet undue optimism about the radical opening up of criminology's research agenda needs to be tempered by awareness of the political constraints which continue to hinder criminological research. For example, research on gun-related crimes committed by the so-called 'underclass' is much more likely to be looked upon favourably by official sponsors than the type of research on state-related misdemeanours and harms discussed by Tombs in Chapter 3.

The example of undertaking research on the police in the 'old' South Africa graphically illustrates the difficulty of starting research in the face of opposition from the state. Commenting on the lack of research on the South African police and their own reliance on secondary data from the media, Brogden and Shearing (1993) show the immense influence of state political power on the nature of the research process. Brogden and Shearing note that in the 'old' South Africa most state practices were defined as beyond legitimate public concern and, thus, beyond academic inquiry. They then go on to note that there was a methodological impasse with regard to obtaining reliable data on the South African police: 'It is no fault of critical South African academics that research on policing in South Africa is notable for its relative barrenness . . . Parliamentary debates have indicated that the line between criticizing the police and subverting the "national interest" has been a very thin one' (Brogden and Shearing, 1993: 192). As a result, Brogden and Shearing had to rely largely on newspaper reports as a major source of illustrative material in putting together their analysis of the rationale for, and operations of, the 'old' South African police force.

It is rare for criminological research in contemporary democracies to encounter such obvious barriers. Nonetheless, extreme examples or *causes célèbres* are significant in highlighting the limits of 'open government' in the criminal justice system. The classic example of state control and censorship of research into the criminal justice system in the UK remains Cohen and Taylor's study of life-sentence prisoners at Durham Prison (Cohen and Taylor, 1972) which is discussed below.

Gaining support and sponsorship

Different theoretical approaches in criminology generate specific research strategies and, in turn, look to different 'sponsors' for support. Nor should we assume that research agendas themselves are created outside 'sponsor' and gatekeeper influence. Rather, the agendas emerge as the by-product of a set of preferences by sponsors for research of a particular character. These preferences generally include a concern for 'relevance' or utility to the policy process (as described by Crow in Chapter 6); individualistic explanations which locate 'pathology' at the level of the individual rather than at the structural level; and finally, quantitative findings which have a seemingly 'hard' scientific basis.

However, given the principle of academic autonomy, researchers do not always 'supply' their sponsors with what they may wish to hear. Not all sponsored research is uncritical of the status quo or subservient to the interests of the sponsor. The research programme of the government-funded Economic and Social Research Council (ESRC) on criminal justice in the 1980s produced critical and systematically sceptical findings on government and Home Office criminal justice policy in virtually all its projects, ranging from studies of penal policy to those of community consultation and policing (Downes, 1992). One illustrative example from the programme clearly highlights the critical possibilities opened up by state funded research. In their study 'Control, security and humane containment in the penal system in England and Wales' (1992), King and McDermott investigated five representative prisons in the Midlands in order to compare current conditions with those in the 1970s. Drawing on such indicators as length of time spent in a cell and amount of time available for work, training and education, the study painted a depressing picture of deterioration in conditions: 'In every comparison much less time was spent in work or similar activities today than had been the case fifteen or more years ago' (King and McDermott, 1992: 104). Their recommendations were critical of current government policy on sentencing and penal policy. King and McDermott recommended a 'minimum-use-of-custody' principle for government and the courts which would bring the prison population into line with existing resources; a 'minimum-use-of-security' principle to alleviate the deterioration of conditions within prison; and, finally, legally enforceable standards of custody to ensure that resources were maintained at levels appropriate for the tasks of the Prison Service.

Although such research stopped short of calling for the abolition of the prison, this example illustrates the relative autonomy of academic-sponsored research from the ideological preoccupations of the dominant authorities. However, as we shall see later, it is quite another thing to argue that such well-informed research-based recommendations will have any noticeable effect on government policy on law and order and the dominant 'prison works' credo.

Having noted the important qualification about academic autonomy and critique, let us now explore how the different means of gaining support may involve different political constraints and opportunities for the researcher. The typical sponsors for criminological research range from direct state sponsorship (for example, the Home Office and Royal Commissions) to relatively autonomous government research councils (for example the ESRC) through to local government support (unitary borough, metropolitan and county councils), private charitable organizations (for example the Joseph Rowntree Foundation), university departments and employer organizations (Bulmer, 1982). Each of these sponsors brings different political agendas to the researcher–sponsor relationship.

Most criminological research may be defined as 'sensitive' research in that it has potentially serious consequences for all participants (Lee, 1993). Much criminological research may also be surrounded by political and moral controversy in that it 'illuminates the dark corners of society' (Lee, 1993: 2). Not least among our society's 'dark corners' are the institutions of the criminal justice system itself, which are often closed and secretive.

The most famous example of criminological research in the UK to come under the withering gaze of this 'secret state' is the previously mentioned research of Cohen and Taylor on life-sentence prisoners in Durham Prison's E Wing (Cohen and Taylor, 1972). This much-discussed project illustrates the highly charged political climate in which penal research has to operate. The research project focused on life-sentence prisoners' 'talk' about how they coped with long-term imprisonment and it clearly departed from the dominant positivist psychological work previously carried out under the banner of 'psychological deterioration'. Cohen and Taylor's research pointed to the brutalizing and dehumanizing effects of institutionalization which are compounded by the tyranny of a sense of endless time, unbroken by the escape routes available to most of us 'outside' the walls of the prison. For the Home Office, such a project was flawed, biased and 'unscientific', given its qualitative methodology, its small sample and its subjective approach. It was also suggested by the Home Office that the research ignored the fact that conditions in prisons were improving (Cohen and Taylor, 1977). At the end of their 'battle' with the Home Office, Cohen and Taylor offered an analysis of its political power. Five sources of power are distinguished by the authors:

1 centralization of power in the prison department;
2 legalization of secrecy, particularly through the Official Secrets Act;
3 standardization of research;
4 mystification of the decision-making structure, given the impenetrability of civil service decision-making;
5 appeal to public interest.

The power of the Home Office to set limiting criteria in the form of a customer–contractor basis for deciding on the agenda of research to be funded has been questioned by Radzinowicz (1994). According to Radzinowicz (1994: 101) this development of a customer–contractor relationship 'can lead to the agenda for research being set in relation to the administrators' (and ultimately the Minister's) conception of what kind of knowledge is needed. Moreover, it will also lead to a preference for short-term inquiries on matters of immediate concern which are regarded as politically or administratively urgent.' The dangers to criminology of this institutional pressure for 'policy-relevant' research are multifold, but, most pressingly, they may lead to a dissolution of genuine analytical questions due to the pressure of delivering what the 'customer' narrowly defines as relevant.

This noted, it would be inaccurate to stress the closed nature of the criminal justice system in the UK. For example, in the early 1990s there was some evidence of the Home Office's willingness to have elements of the system monitored. In part, this was due to the legal requirements of the Criminal Justice Act of 1991 whereby race and gender monitoring of the criminal justice system was required to be published annually and this partial opening up of the system to research also arose out of a concern for the legitimacy which external 'audits' (to use the language of 'new managerialism') offered public services. It is debatable how influential independent studies are on actual policy and practice in the Prison Service, but the partial opening up of the system to research or audit does show that the Home Office is not a monolithic bastion whose every gate is forever closed to critical academic scrutiny.

Gaining access

In order to carry out social scientific research it is often necessary to get into an institution or an informal grouping to make first-hand contact with the subjects of study. The problem of access does not end, however, once 'in' through the door. Gaining access is an ongoing process of negotiation and renegotiation. In particular, a crucial role is played by key individuals (termed 'gatekeepers' in the academic literature). Gatekeepers may be defined as those individuals in an organization or another social situation who have the power to grant or withhold access to people or situations for the purposes of research. Furthermore, such gatekeepers are not necessarily formally 'in charge of' a given institution or group. As with many sites of research, there are often multiple points of entry into an institution or social setting. For example, any research into the existence of an informal subculture among police officers would need to have access to the rank-and-file 'canteen cowboys' rather than the formal policy statements of senior police managers if it is to explore the issue in an adequate fashion (Smith and Gray, 1983).

It is also crucial to note the differences of degree in the relative openness of access between different groups, which in turn reflects the differential degrees of power of those being researched (to take two examples of starkly different phenomena, compare the ease of access to information about the incidence of criminality among London's homeless youth as against that of the unreported crimes of City financiers). 'Studying down' (that is, studying vulnerable minority groups) is much easier and commonplace in criminology than is 'studying up' of powerful elites. Homeless youth is subject to much more public scrutiny than City financiers, whether we are referring to the policing of their behaviours, to media reportage or, indeed, to academic research. The young homeless as a vulnerable group also lack formal protection from academic scrutiny. It would be quite straightforward to investigate known offenders among homeless youth with likely high levels of subsistence crime and 'disorderly' behaviour, such as shoplifting, vagrancy and begging. In contrast, City financiers possess a great many resources for maintaining their privacy and freedom from intrusion from academic researchers who may wish to study such crime as unreported fraud and illegal dealing. Closed doors, many gatekeepers and a privileged culture of privacy will confront any researcher who wishes to study business-suite crime. Undercover ethnographic research would be possible but very difficult given the technical knowledge required of any researcher who hopes to overcome the problem of 'passing' as a financier.

The gaining of access may be either highly procedurized or personalized (Lee, 1993: 124). An example of a highly procedurized relationship is the conditional access associated with conditions laid down by the Official Secrets Act in the UK. Much of the research undertaken for the Home Office Police Research Group, for example, would fall into this category of procedurized access, with the Home Office maintaining the right to examine, modify or block any material to be published. In contrast, a more personalized mode of access often involves a designated 'chaperone' to escort the researcher down the metaphorical 'corridors' of the research process. Such access may also be very restricted in that despite this apparent openness, in reality the researcher is sent down wrong corridors into culs-de-sac and/or the personalized chaperone controls the nature of the information gathered.

There is likely to be an imbalance of power between gatekeeper and researcher which will lead to bargaining between the two parties: the so-called 'research bargain'. If research is viewed as potentially threatening, the process may generate what is often termed 'the politics of distrust' whereby each side is suspicious and secretive about the activities of the other. Perhaps unsurprisingly, gatekeepers tend to prefer methods which are thought to deliver 'hard facts' and which as a result offer the gatekeeper some scope for scrutiny and control: this explains the attraction for gatekeepers and sponsors of quantitative surveys and questionnaires rather than qualitative observation and interviews (Lee, 1993: 124). This is illustrated by research into a multi-agency diversion and crime prevention unit in Northamptonshire (Hughes et al., 1998) which involved quite sensitive and strained negotiations and bargaining about the balance of detailed qualitative interviews and observation to explore the hidden history of the unit in question (the research team's main interest) and the senior management of the unit's wish to get seemingly 'hard' quantitative data showing the 'success' of the Diversion Unit in reducing reoffending. There is also a hierarchy of consent in all formal organizations. It would be very dangerous for researchers to assume that 'superiors' (that is, formal gatekeepers) have the right to allow 'subordinates' to be investigated. In reality it is not uncommon for researchers to get formal physical access without the accompanying informal social access. This may be termed 'the micro-politics of research'. This situation should alert us to the significance of informal gatekeepers who may erect unofficial barriers to the research process. To cite the example of Magee and Brewer's ethnographic research on the RUC (Brewer, 1991), access to the force was granted eventually by senior officers since allowing such research to take place would be a good public relations exercise and would show that the RUC was open to public scrutiny. There was thus a pay-off for senior echelons of the RUC in terms of their professional ethos and legitimacy. However, it was likely that the research would be less popular with the rank-and-file police officers who ran the risks of answering awkward questions and being observed doing their often 'messy' work. Having dealt with the first gate of the senior 'gatekeepers', Magee had a much more difficult task gaining 'social' access through the informal gate of the rank and file. Gaining the trust of this group was to prove a long-drawn-out process, in Brewer's words 'a result of a progressive series of negotiations . . . continually negotiated' (Brewer, 1991: 19).

Examples from police research studies illustrate that it is very difficult to negotiate successful access to the routine operation of the criminal justice system and its practitioners. Nevertheless, the growing body of research on the rank-and-file police officer has uncovered some key findings with regard to such phenomena as the role of discretion in routine police decision-making and the importance of the informal occupational 'cop culture' in moulding rank-and-file attitudes and behaviour. However, very little is known of the life of the men and the (few) women at the top of the police force. Access to elites in the criminal justice system is rare for researchers. An exception to this rule is Reiner's study of chief constables in England and Wales (Reiner, 1991). It would be a salutary lesson to any budding 'elite studies' researcher to read Reiner's frank reflections on the attenuated process of negotiating access to the chief constables which accompanied his research proposal (see Reiner, 1991: Chapter 3).

It is also highly probable that researchers are 'checked out' by information-gathering institutions such as the police in terms of their previous work and ideological leaning. In the era of 'late modernity', the institutions of the criminal justice system are likely to be knowledgeable and reflexive and quite often the researcher's reputation will precede him or her. This may create complex problems for the researcher in 'passing over' as a legitimate person. There is the likelihood that he or she will be tested and face unofficial rites of passage. As a researcher, it is not uncommon to hear stories from the researched about the character and outcomes of previous research. One example involves comments from senior police officers with regard to a researcher studying victims' and offenders' treatment by the police who is now commonly known in this particular police force as the 'study and snitch' researcher since she seemed to them friendly during the research and then produced what they thought was a damning written report. Whatever the rights and wrongs of this case, it is a useful illustration of the type of informal barriers to access which may be erected by institutions following earlier research experiences.

Collecting data

The above example of a sense of betrayal felt by the researched leads to the next stage of the research process, namely the gathering of data, and also to the issue of the extent to which 'informed consent' is required for ethically sound research. It will be evident that the collection of data is far from a purely technical exercise, but is itself a form of political activity. Most research is carried out on the relatively powerless. Exposé research of dominant elites and institutions is the exception rather than the rule, but it is alive and well in critical criminology. For example, Hudson argues that one of the defining characteristics of critical social science is that it works 'on behalf of those on the downside of power relations' (Hudson, 1993: 7), a theme which she reiterates in Chapter 10 of the present book. Worrall's research on the regu-latory discourses surrounding female law-breakers also deliberately avoided asking such questions as 'why certain women offend', since this might perpetuate dominant practices and ideologies with regard to what Worrall argues are oppressed women. Instead of retreading the traditional path of searching for causation and aetiology, Worrall, in collecting her data, sought to 'examine the ways in which the authorization of professionals and experts to define certain women as being the type of woman who requires treatment, management, control, or punishment serves to perpetuate the oppression of all women' (Worrall, 1990: 4). This was done by studying 15 female law-breakers, together with 'experts' such as magistrates, solicitors, psychiatrists and probation officers.

In contrast to such critical research, but in other ways similarly 'partisan' and involved, is research which is fully supported by the agency being investigated. Waddington's (1991) research into armed weapons training and public order policing policies of the Metropolitan Police offers an interesting illustration of 'criminology from above'. Waddington, himself an ex-police officer, appears very much an 'insider' with easy access to the normally hidden world, receiving, as he notes, 'the fullest co-operation from . . . all ranks of the Metropolitan Police' (Waddington, 1991: 271). There is much to recommend in this study 'from the inside' of changing police policy and practice, given its privileged access to gathering data rarely available to the outside researcher. At the same time, questions may be raised as to the detachment and objectivity of a researcher who becomes very closely involved with, and possibly on the side of, the research participants. In the course of his overt participant observational research, Waddington arrives at the conclusion that there are distinct advantages both for the police and citizens in the existence of a professional and specialized paramilitary policing strategy.

The question of 'whose side are we on?' as criminological researchers is unlikely to be resolved given the plurality of approaches in current criminology. Traditionally, much mainstream criminology has been on the side of the criminal justice system and much research continues to adopt this (often) unstated political position. However, it is a sign of criminology's vibrancy that the question of allegiance is now openly aired and debated.

That an individual should be informed of the nature of research in which he or she is a subject, and its likely dissemination, may seem an undeniable right to which any human is entitled. Indeed, in most of the social sciences there are codes of practice which seek to establish this entitlement as part of the working philosophy of researchers. The value of such guidelines is that they acknowledge explicitly the power relations between researcher and researched. They are particularly helpful in protecting the vulnerable from exploitation by researchers. Furthermore, they have the ethically powerful appeal of being open and honest about the actions and outcomes of the work undertaken by the researcher. It is likely that research conducted along these lines will maintain a reasonable level of public trust. However, critics of such guidelines point to the danger that research organized on the basis of informed consent may not uncover important data since the co-operation of the research participants should be sought and those with things to hide may be unwilling to allow research to be undertaken! This would limit the capacity for exposé research. More generally, it might be argued that life itself involves lies and distrust and, as a consequence, such codes are a denial of reality.

Clearly not all research in criminology has been characterized by openness and informed consent in the manner of its data collection. Covert or undercover research negates the principle of informed consent, as those being researched cannot refuse involvement. Of necessity, certain research needs to be covert. This claim applies in particular to research on the powerful and the privileged. Overt research, according to this argument, favours the rich and the powerful. As Benyon has remarked: 'Historically the rich and powerful have encouraged hagiography, not critical investigation' (quoted in Lee, 1993: 8).

Where research is threatening, the relationship between the researcher and the researched is likely to become hedged about with mistrust, concealment and dissimulation (Lee, 1993: 2). This viewpoint is relevant to studying not just the very powerful but any group with vested interests, for example bakery workers and their perks (Ditton, 1977). Ditton is not concerned by the accusation of subterfuge with regard to his own undercover gathering of data on the routine fiddling in the 'Wellbread' bakery: 'Without reliance on some subterfuge the practices of subterfuge will not be opened to analysis' (Ditton, 1977: 10). However, other commentators *are* concerned about the ethics of such deceitful research. Criminologists are thus faced with a difficult balancing act between the quest for greater human knowledge and harm done to individuals in the pursuit of this goal. For philosophers this represents the classic means/ends dilemma for which there is no easy answer.

There is a powerful argument that researchers need to compromise or we end up seeking 'to understand how angels behave in paradise' (Klockers quoted in Lee, 1993: 139). Such compromises may involve the criminological researcher in complicity in wrongdoing – for example, the witnessing of malpractice among the police, witnessing illegalities whilst being a participant observer of juvenile delinquency or being a non-participant covert observer of drug dealing. Some accommodation to, and appreciation of, the world of the deviant/wrongdoer may be necessary to the successful gathering of data in some criminological research.

Publishing the results

When research findings are published criminologists enter another political arena in which the researcher has to take account of a variety of audiences including research participants, sponsors, funders, the public, and other academics. It is highly probable that all these audiences have distinct, differing and possibly conflicting expectations, and that there will be people who feel damaged and threatened by the publication of criminological research findings. On occasions criminological researchers themselves may be threatened as a result of their conclusions and policy recommendations which question certain vested interests. For example, Walters' (1997) research on gun control in New Zealand resulted in him receiving hate mail and telephone calls from members of the pro-gun lobby.

It has already been noted that criminological research may be subject to state censorship. In the past research has occasionally been subject to legal writs regarding defamation. In the British context, Baldwin and McConville's research into plea-bargaining (1977) remains the most famous example of research which met pressure for censorship of its findings. Baldwin and McConville's research uncovered worryingly high rates of informal bargaining and negotiation as to the plea despite the formal denial in law that such practices existed at that time in England's adversarial system of justice. Evidence was found of defendants being persuaded to change their plea from innocent to guilty as a result of pressure, often from their own legal representatives. The project unearthed evidence that due process and adversarial justice were routinely undermined in practice in the Crown Courts. There thus appeared to be a gulf between what is often called the 'law in books' and the 'law in action'.

Baldwin and McConville's *Negotiated Justice* remains an instructive – if extreme – exemplar of the politics of publishing. Prior to its publication, the authors were confronted with a public controversy emanating from the Senate of the Inns of Court and representatives of the Bar. The legal establishment in England was set against the publication of the study. Only after a prolonged series of negotiations in which external academic consultants were commissioned to check the validity of the research findings did the research get the go-ahead for publication. In the longer term, it is worth pointing out the undeniable value of the work in opening up the criminal justice system to critical scrutiny and more open accountability.

The utilization of criminological research by policy-makers

Most commentators on this issue accept that the effects of social sciences are often negligible in the sense of exciting public interest. That noted, the highest-profile utilization (and non-utilization) of criminological research is that associated with royal and other government commissions in the UK.

Routinely, commissions collect evidence, analyse the problem, report publicly and make recommendations for governmental action. In arriving at their conclusions, Royal Commissions on Criminal Justice do draw on criminological research evidence, although the extent of academic research influence is a matter of some controversy. It is often argued that the possibility of a consensus at times appears to be a more important factor than the most accurate explanation of events. There is then a long history of political neglect of research-based policy recommendations. The 1993 Royal Commission on Criminal Justice (RCCJ), or the Runciman Commission, so named after its chair, is a useful illustration of this point. To put the latest RCCJ in its political context, the Home Secretary announced the establishment of a RCCJ in 1991 on the day when the 'Birmingham Six' – six individuals wrongly convicted for terrorist bombings – were released after 16 years of imprisonment. The brief of the Commission seemed to be to address the many problems associated with unsafe convictions and there is little doubt that its setting up was an exercise in addressing a crisis of legitimacy in the criminal justice system. When eventually established, the terms of reference of the RCCJ were to examine the workings of the system from the stage at which police investigations of alleged criminal offences occur right through to the point at which the defendant has exhausted his or her rights of appeal. Furthermore, the emphasis moved from that of wrongful convictions to encompass effectiveness in securing convictions and the efficient use of resources (Field and Thomas, 1994: 2).

To assist the Commission in its work, a programme of academic research was commissioned and organized by the Home Office Research and Statistics Department. When the Commission came to make its 352 recommendations the response from much of the academic community was anger and disappointment at the neglect of its research. Viewed critically, the RCCJ has been termed a commission which 'normalizes injustice' (Bridges, 1994). In order to understand the controversy over the apparent non-utilization of academic research by the RCCJ, some of its key recommendations need to be outlined.

The RCCJ rejected most of the academic suggestions put to it which were aimed at reducing the likelihood that the innocent would be convicted. That noted, the RCCJ did make some important recommendations with regard to the 'efficiency' of the criminal justice system such as that, in cases of either-way offences, the defendant should no longer have a right to trial by jury. Pre-trial disclosure of defence evidence was extended and plea-bargaining was more overtly accepted with a higher discount for an early plea. More scope for procedures which clarified issues before trial was also recommended. As Field and Thomas (1994: 4) note, the Commission's most radical proposals went in the opposite direction, of greater safeguards against miscarriages of justice because of its focus on greater efficiency. Higher standards for both police and defence lawyers were recommended and an end to the right to silence was rejected.

Following the publication of its report, the utilization by the government of the RCCJ's recommendations was highly selective and this illustrates the complex ways in which research and its policy recommendations are liable to be refracted, if not ignored, by dominant political discourses. Addressing the Conservative Party Conference on 6 October 1993, the Home Secretary, Michael Howard, announced the unveiling of a dramatic and draconian 'law and order package'. As Field and Thomas (1994: 7) again note:

> All the penal lessons that had been painfully learnt in the early 80s were lost in a scramble for a law and order rhetoric with popular appeal. Inconvenient evidence was simply ignored. In this process, the Runciman Commission was cannibalized: anything that could be presented as a contribution to cost-effective crime control became an urgent political priority. Other issues could wait. Some of the announced changes follow the Runciman Report closely; others flatly reject the majority view or simply cut across the Commission's assumptions.

The broad message seems to be that governments are able to pick and choose which pieces of informed recommendations to accept and which to banish from the populist 'law and order' discourse, or discreetly shelve.

From the above outline of the history of the 1993 RCCJ, it should be clear that politics plays a central and determining role in the ways in which research will be utilized, or not as the case may be, in policy-making. During this same period of Michael Howard's tenure as Home Secretary, it appeared that ministers were preventing publication of evaluation research from within the Home Office which contradicted Howard's proclaimed 'law and order crackdown' and 'prisons work' credo (*Guardian*, 4 July 1994). It is evident that other actors, including pressure groups, in the 'crime control industry' also use research findings selectively. Accordingly NACRO (National Association for the Care and Rehabilitation of Offenders), formerly a radical, left of centre pressure group, together with the Audit Commission (1996), has used research on multi-agency crime prevention in a selective manner in order to make its case for reparation and multi-agency crime prevention partnerships to the New Labour government as an alternative to more traditional approaches to dealing with offending. For example, the article 'Diversion tactics' by NACRO's Lynne Wallis in the *Guardian* Society of 26 August 1998 paints a simplified message of the original research study's tentative and very qualified findings on the effects of this Diversion Unit in reducing reoffending (see Hughes et al., 1996). Selective use of research findings is therefore not the preserve of right-wing politicians but instead characterizes much of the policy-making politics in the crime control 'business'.

Conclusion

This chapter has explored the realities of doing criminological research in what is a highly politicized world. Criminological research does not occur in a metaphorical germ-free, antiseptic zone. We need to be wary of the talk of the end of politics and rise of non-political technical fixes for research. Such talk is likely to usher in very restrictive research agendas for criminology. Furthermore, it is impossible to envisage a time when criminological research will not generate the types of political controversies and ethical dilemmas discussed in this chapter.

The long-term influence of research on both the policy and political process should not be lost. Indeed criminological research over time, and as a result of its capacity to generate contestation and dialogue, may help redraw the frameworks employed in political discourses on law and order.

Suggested readings

Hammersley, M. (1993) *Social Research: Philosophy, Politics and Practice*. London: Sage.

Jupp, V. (1989) *Methods of Criminological Research*. London: Routledge.

Lee, R.M. (1993) *Doing Research on Sensitive Topics*. London: Sage.

May, T. (1997) *Social Research*, 2nd edition. Buckingham: Open University Press.

References

Audit Commission (1996) *Misspent Youth*. London: Audit Commission.

Baldwin, J. and McConville, M. (1977) *Negotiated Justice*. Oxford: Martin Robertson.

Brewer, J. with Magee, K. (1991) *Inside the RUC: Routine Policing in a Divided Society*. Oxford: Oxford University Press.

Bridges, L. (1994) 'Normalizing Injustice', *Journal of Law and Society*, 21 (1): 20–34.

Brogden, M. and Shearing, C. (1993) *Policing for a New South Africa*. London: Routledge.

Bulmer, M. (1982) *The Uses of Social Research*. London: Allen and Unwin.

Cohen, S. and Taylor, L. (1972) *Psychological Survival*. Harmondsworth: Penguin.

Cohen, S. and Taylor, L. (1977) 'Talking about prison blues', in C. Bell and H. Newby (eds) *Doing Sociological Research*. London: Allen and Unwin.

Ditton, J. (1977) *Part-time Crime*. London: Macmillan.

Downes, D. (ed.) (1992) *Unravelling Criminal Justice: Eleven British Studies*. Basingstoke: Macmillan.

Field, S. and Thomas, P. (1994) 'Justice and efficiency? The Royal Commission on Criminal Justice', *Journal of Law and Society*, 21 (1): 1–19.

Hudson, B. (1993) *Penal Policy and Social Justice*. London: Macmillan.

Hughes, G. (1998) *Understanding Crime Prevention: Social Control, Risk and Late Modernity*. Buckingham: Open University Press.

Hughes, G., Leisten, R. and Pilkington, A. (1996) *An Independent Evaluation of the Northamptonshire Diversion Unit*. Northampton: Nene Centre for Research.

Hughes, G., Leisten, R. and Pilkington, A. (1998) 'Diversion in a culture of severity', *Howard Journal of Criminal Justice*, 37 (1): 16–33.

King, M. and McDermott, K. (1992) 'Control, security and humane containment in the penal system in England and Wales', in D. Downes (ed.) *Unravelling Criminal Justice: Eleven British Studies*. Basingstoke: Macmillan.

Lee, R.M. (1993) *Doing Research on Sensitive Topics*. London: Sage.

Radzinowicz, L. (1994) 'Reflections on the state of criminology', *British Journal of Criminology*, 34 (2): 99–104.

Reiner, R. (1991) *Chief Constables*. Oxford: Oxford University Press.

Smith, D. and Gray, J. (1983) *The Police and the People in London*. London: Policy Studies Institute.

Waddington, P. (1991) *The Strong Arm of the Law*. Oxford: Clarendon Press.

Walklate, S. (1998) *Understanding Criminology*. Buckingham: Open University Press.

Walters, R. (1997) 'Gun control in New Zealand'. Paper delivered at Australian and New Zealand Criminology Conference, Wellington, July.

Worrall, A. (1990) *Offending Women: Female Lawbreakers and the Criminal Justice System*. London: Routledge.

INDEX